KGB

ALPHA TEAM TRAINING MANUAL

How the Soviets Trained for Personal Combat, Assassination, and Subversion

PALADIN PRESS
BOULDER, COLORADO

KGB Alpha Team Training Manual:
How the Soviets Trained for Personal Combat,
Assassination, and Subversion

Copyright © 1993 by Paladin Press

ISBN 10: 0-87364-706-8
ISBN 13: 978-0-87364-706-9

Printed in the United States of America

Published by Paladin Press, a division of
Paladin Enterprises, Inc.,
Gunbarrel Tech Center
7077 Winchester Circle
Boulder, Colorado 80301 USA
+1.303.443.7250

Direct inquiries and/or orders to the above address.

PALADIN, PALADIN PRESS, and the "horse head" design are trademarks belonging to Paladin Enterprises and registered in United States Patent and Trademark Office.

All rights reserved. Except for use in a review, no portion of this book may be reproduced in any form without the express written permission of the publisher.

Neither the author nor the publisher assumes any responsibility for the use or misuse of information contained in this book.

Visit our Web site at www.paladin-press.com

TABLE OF CONTENTS

Preface 1
Translator's Note 61
Foreword 65

Chapter 1. The Foundations of Special Physical Training ...67
    Foundations of Special Physical Training Organization
    Principles of Instruction

**Chapter 2. Movement; Overcoming Obstacles;
Penetrating/Storming Buildings 87**
    General Systematic Directions for Teaching
        Basic Methods of Movement
    Fundamentals of Movement and Overcoming Obstacles
    Movement under Special Conditions
    Special Features of Night Movement
    Leaping Natural Obstacles
    Running and Crawling
    Movement in Mountains
    Movement in Deserts
    Overcoming Man-Made Obstacles and Positions
    Crossing Water Barriers

**Chapter 3. Techniques and Methods for
Teaching Personal Combat** 123
  Recommendations for Methods in Teaching Tactics of
    Personal Combat
  Basic Vulnerable Areas and Points of the Human Body
  Techniques of Inflicting Effective Blows
  A Graduated Series of Warm-Up Exercises
  Special Exercises
  Blows
  Safety and Self-Protection in Falls
  Self-Protection in Falls to the Side
  Safety in Forward Falls
  A Series of Exercises in Learning Safety/Self-Protection

**Chapter 4. A Practical Section in
Special Physical Training** 147
  Basic Methods for Capturing
  Basic Methods for Silently Killing an Armed Enemy
  Additional Methods for Silent Killing
  Cold Weapons
  Choking Techniques
  Attacks by Teams
  Silent Attacks on the Enemy from Concealment
  Attacking an Enemy in Its Position
  Capturing an Enemy Traveling by Bicycle,
    Motorcycle, or Horse
  Signs and Signals for Silent Operations
  Some Training Exercises and Tasks
  Methods for Securing and Transporting Prisoners
  Methods for Securing
  The Use of Handcuffs for Securing
  Methods of Conveying a Prisoner
  Methods for Evacuating the Wounded

**Chapter 5. Escaping from and Fighting Off
Physical Attacks; Mutual Aid; Throws** 191
  Escaping Attacks from the Front

TABLE OF CONTENTS

Escaping Attacks from Behind
Escaping from Holds in Fights on the Ground
Defense and Mutual Aid
Self-Defense against an Enemy with a Firearm
Basic Methods of Defense against a
   Firearm Aimed from in Front
Basic Methods of Defense against a
   Firearm Aimed from Behind
Defense against Cold Weapons
The Overhand Arm Knot Lock
Inward Arm Twist against an Overhand Stab
Underhand Stabs
Backhand and Lateral Stabs
Self-Defense Using Additional Means
Basic Techniques
Additional Ways to Defeat an Enemy without Using Weapons
Twisting the Neck Vertebrae
Choking Techniques
Using Weapons and Other Objects for Self-Defense
Throwing Cold Weapons at a Target

Chapter 6. Penetrating Buildings in an Attack.................277

Chapter 7. Models for Restoring Work Capacity and Monitoring the State of Health 285
Steam Baths
Nutrition in Times of Heavy Physical Exertion
Vitamins
Water
Ways of Monitoring the State of Health
Some Possible Breakdowns in Human Health under
   Heavy Stress
Injuries

Readings...311

WARNING

The information presented in this book is *for reference and historical purposes only*! The author, publisher, and distributors do not in any way endorse nor condone any illegal or dangerous activity or act that may be depicted in the following pages. Therefore, the author, publisher, and distributors disclaim *any* liability and assume no responsibility for the use or misuse of the information herein.

PREFACE

Editor's note: The KGB Alpha Team Training Manual *was provided to Paladin Press by Jim Shortt, who, as director of International Bodyguard Association (IBA), has trained numerous Western military and police units in anti-Spetsnaz activities. Shortt was the first outsider to train KGB personnel, and he has been active in the Baltic States both before and after independence, training these republics' police and security forces. Shortt also trained mujahideen forces during the war in Afghanistan.*

Several pages in chapters 5 and 6 of this manual are missing. The same pages were missing in every copy of the manual that Shortt examined. This leads one to believe that the pages were either deliberately pulled because of sensitive information found on them, or the Soviet military suffered from the same inefficiency as bureaucracies everywhere and the pages were inadvertently left out of the original printing. The places with missing text have been footnoted.

In the following, Shortt briefly examines Soviet special operations to show the relationship of various organizations and to document how the information contained in the manual was used by the KGB, GRU, MVD, and other "special assignment units." He also includes some personal accounts of his training missions in various Soviet republics to illustrate how many of the functions formerly performed by the KGB and GRU are now being assumed by police units in the various republics or local mafia groups.

I was sitting in a small apartment in the Latvian capital of Riga in January 1992 with members of the Latvian Security Service's bodyguard department. Between us we were—as the Irish in me would say—doing justice to a goodly number of bottles of *Kristal Dzidrai*s, Latvian vodka, and *melnais balzams*, a potent tarlike local liquor. Our host, a major with the service, had been in his time a graduate and later instructor at the Soviet Defense Intelligence (GRU) #4 Spetsnaz[1] Brigade based near the Estonian town of Viljandi.

While the snow and minus-16-degree temperature kept the Latvian vodka-in-waiting correctly chilled, I pored over the photograph album of my host and mused that it was, in many ways, similar to my own. Although the uniforms and equipment were different, the scenarios were similar. When I came to the training manuals used by the Soviet Spetsnaz, I noticed that they were surprisingly few and all written in 1945 by veterans of the partisan units, OSNAZ Brigade, and Reconnaissance Scouts. Their primary emphasis was on physical capability, daring, and conditioning.

Next, I looked over more recently produced close-quarter-battle (CQB) manuals from the army physical training department and the Naval Infantry,[2] termed in Russian *rukopashnyi boi*. They covered unarmed scenarios, edged weapons (such as the bayonet, entrenching tool, and knife), and finally projectiles, as well as the techniques for throwing bayonet, rifle and bayonet, entrenching tool, and a special sharpened steel plate. Just when I thought I had seen it all on special combat techniques!

The manual you now hold in your hands has been translated from its original Cyrillic format. I was told that it was a very special manual because it was produced by A.I. Dolmatov, the man who had trained the KGB special units codenamed "Alpha" teams at the Moscow Dynamo sports club. If you had asked any official of the former Union of Soviet Socialist Republics (USSR) what Dynamo was, he would have answered, "Just a sports club; policemen use it." Ask any Western sovietologist the same question, and he would add that the Dynamo organization existed in every major Soviet city and was dominated by the KGB and the MVD (Soviet Interior Ministry). It was not your usual after-work squash facility, but rather an integral part of the training and update of the Soviet regime's countersubversion forces. This manual was produced for the special forces of the MVD[3] and the KGB,[4] as well as Defense Ministry personnel seconded to them.

This manual was intended for the training of personnel operating on internal security duties within the Soviet Union and also in in-depth missions against enemies of the Soviet Union. Soviet Spetsnaz troops operated from front lines of battle up to 1,000 kilometers to the enemy's rear.

The Interior Ministry controlled two types of personnel: the MVD militia, or Soviet police, and the MVD (VV), or Internal Forces. The MVD militia had their special forces in the OMON formations while the MVD units—which were the de facto internal army of the Soviet Union—had specialist units called Spetsnaz Soviets.

The task of the internal army was putting down rebellion and hunting Western special forces that landed in time of war behind Soviet lines. Sandwiching the interior army of the MVD and the Defense Ministry's exterior army was the KGB—its First Chief Directorate that had the exterior army was the "Cascade" (*Kaskad*) program for offensive special forces operations against the West, including assassination and sabotage. The Second Chief Directorate with the Chief Directorate of Border Guards that had control of special units within the Soviet Union, especially the KGB Alpha teams that cross-trained for the Cascade program. To understand the different types of Soviet special assignment forces that existed (and still exist to a large extent within the Confederation of Independent States [CIS], the successor to the USSR), it is necessary to examine the development and evolution of such units from the Bolshevik seizure of power in 1917.

HISTORY OF SPECIAL OPERATIONS

The Bolshevik Revolution was not in fact a Russian revolution. Lenin maintained himself in power by force of arms. First he used Latvian infantry to guard the Kremlin in Moscow against the Russian people, and second he appointed a Pole, Felix Dzerzhinsky, to be in charge of state security. On 20 December 1917, the All-Union Supreme Commission to Combat Counterrevolution, Sabotage, and Speculation was set up under Dzerzhinsky; it was known by the abbreviation VChK or Cheka. It was to the Communist party what the SS was to the Nazis. The Cheka command structure held no Russians, just international Communists who were Czechs, Latvians, Austrians, Poles, Hungarians, Finns, and other non-Russians. The VChK would subsequently become known as the GPU, OGPU, GUGB, and then finally the NKVD (The People's Commissariat for Internal Affairs).

Stalin formed special units to carry out assassinations abroad (of rivals such as Trotsky in Mexico) and to rid Stalin of internal opponents and those who did not actively support him. In 1936, the Cheka created an Administration for Special Tasks to kill or kidnap persons outside of the territory of the USSR who were deemed enemies of the state. However, in mid-1919 the Cheka had already created its first special-operations units, the CHON[5] and later (as the GPU) the elite Dzerzhinsky Division, which, with the break up of the NKVD, became part of the MVD.

In June 1941 when Hitler invaded the Soviet Union, large numbers of NKVD border guards fought against the Nazis and, as Communist party faithfuls, were among the first partisan units operating behind German lines. The NKVD formed a partisan training program at Tiflis, which they christened the "00" program, and the NKVD border guards formed the core of the first NKVD special operations units called *istrebitel'nye batal'ony*, which operated on sabotage missions behind German lines. Soon after the invasion, special NKVD Unit #10 gained control of partisan activity.

The NKVD internal forces formed 15 divisions, which though sometimes committed to front-line fighting were normally used at the rear of the Red Army to prevent retreat or desertion. They were also used to punish populations that collaborated with the Germans. By the close of World War II, the NKVD had 53 NKVD divisions and 28 NKVD specialized brigades in addition to its border-guard units. They fought antiguerrilla actions in Ukraine and the Baltic States.[6] They also carried out political "cleansing" operations, deporting and murdering whole communities whose loyalty to the Communist Party was suspect. During World War II, the NKVD created a special operations brigade, OMSBON.[7] Its members were not called Spetsnaz but rather Osnaz.[8] I have found the term Osnaz applied to designate special purpose units of political origin (i.e., KGB, NKVD, MVD), whereas Spetsnaz is used to designate a tactical or strategic unit of politically reliable personnel. Osnaz are politically superior in role to Spetsnaz. OMSBON had roles both behind German and Soviet lines. It launched 212 units behind German lines—a total of more than 7,000 men. But it also operated against Ukrainian and Baltic States nationalists in hunter teams and extermination squads. OMSBON alone boasted a head count of 140,000 people it had killed. The NKVD ran Osnaz teams in to northern Norway in opposition and duplication to Spetsnaz teams operated by the Soviet Naval Infantry during the German occupation of Norway.

The Soviet army created its own Spetsnaz teams of *razvedchiki* or reconnaissance scouts responsible for diversionary reconnaissance, which meant gathering information by penetrating behind enemy lines, intercepting communications, taking and interrogating prisoners—all while they were there murdering senior officers, and destroying headquarters, weapon dumps, stores, roads, bridges, etc. The Naval Infantry followed the army's example and created its own *razvedchik* units.

THE KGB

The KGB was formed in March 1954. The Central Committee of the CPSU[9] split the NKVD into two distinct organizations. Simply put, this was a security measure by one part of the central committee to prevent a state security body from ever wielding the type of concentrated power the NKVD had exerted under and on behalf of Stalin.

Many figures in the central committee of that period ended up arrested, tortured, and even murdered by the NKVD. The concept behind bisecting the NKVD was to return state security from being the watchdog of the Central Committee to being its lapdog

KGB Alpha Team officer with prisoner. Photo courtesy of Novosti Press Agency

and, sometimes, guardian. From the NKVD were created the KGB and MVD, one to supposedly watch the other.

The MVD took responsibility for the militia or Soviet police force and for the vast internal army, including OMSBON units such as the *zagraditel'nye otryady,* or blocking battalions of the NKVD, which were placed behind Soviet army combat units to prevent retreat and desertions, and also the *istrebitelnye otryady,* or NKVD hunter battalions used to find and liquidate anti-Soviet guerrillas. The MVD also assumed responsibility from the NKVD for the guarding and security of more than a thousand prison camps (gulags).

PREFACE

KGB border guards. Photo courtesy of Jim Shortt

KGB Hunter Groups mounted (above) and on foot (right). Photos courtesy of Jim Shortt

The KGB, through its First Chief Directorate, took responsibility for espionage and, through its Second Chief Directorate, for countersubversion and counterintelligence in the civil population. SMERSH,[10] founded in 1943 as military counterintelligence, became the KGB's Third Chief Directorate. The NKVD's Border Guard units came under the control of the KGB's Chief Directorate of Border Guards. The NKVD's Administration for Special Tasks became the Partisan Fourth Directorate in 1941; in 1946 it evolved into Special Office 1 and later Department 13 of the First Chief Directorate. It later became an independent Department V under the direct control of the KGB chairman and was reserved, euphemistically, for "Central Committee special tasks" only.

SPECIAL ASSIGNMENT UNITS

With the division of the NKVD, its various special assignment units landed either under MVD or KGB control, depending on their roles. This meant that three distinct types of special purpose personnel were available for mission direction within the USSR. Spetsnaz from the Defense Ministry's GRU units, Spetsnazovtsy from the MVD, and Osnaz from both MVD and KGB. Except for the officers, the vast majority of personnel who serve in these special units are conscript servicemen. To assist you in better understanding this manual, I should explain how all Soviet young men have been educated from childhood through a series of military and CPSU-sponsored training programs for their roles in the military.

PREMILITARY TRAINING

It is fair to say that the average Soviet conscript inducted into special assignment units within the GRU,[11] MVD, or KGB began his premilitary training at the age of 10 in an obligatory school program sponsored by the Ministry of Defense called the GTO.[12] Although civilian in nature, this school program was aimed at creating and maintaining a high standard of physical fitness for males and females. Overseen and inspected by the Ministry of Defense's Department of Preliminary Military Training, it was established not only in schools, but also in factories, colleges, and collective farms, and also encompassed some postmilitary service training up to the age of 60 under separate schemes. The GTO program had by three subdivisions in schools:

Age Group	Program Name
10-13	Courage and Skill
14-15	Young Sportsman
16-18	Strength and Courage

These program were introduced in 1967 when conscript service was reduced from three to two years in the hope that part of the time lost to military service would be recouped by this schooltime preparation for service.

Local military units provided the program's instructors, and the final objective was to prepare the boys for conscript service with the defense forces, internal forces, or KGB forces. Under the 1967 Law of Universal Military Service, young men from the age of 18 are required to report for military service. It is usual that GTO instructors organize an additional 80 hours of intensive preinduction training course covering nuclear, biological, and chemical defense; forced marches; martial arts; ski races, cross-country races, and orienteering.

PREFACE

However, in addition to the compulsory GTO program there exists also a voluntary military program run by DOSAAF,[13] which is under the direct control of the Defense Ministry. From the age of 14, children can begin training with DOSAAF. The 1972 DOSAAF regulations state that "the society will provide leadership for the development of military-technical skills." All parachute and flight training in the Soviet Union is under the control of the Defense Forces. The basic training comprises a minimum of 140 hours plus training camps over a period of two years. The youngsters can qualify as pilots and parachutists while also learning to drive and maintain vehicles.

In June 1991, I visited the central military bookshop in Moscow and purchased a number of posters illustrating the workings of Soviet weapon systems from the AK-74 rifle to BMD-1 tanks all published by DOSAAF. I also purchased a copy of *Kniga Yunnogo Armeetsa*, the young soldier's handbook published in 1989 by DOSAAF and aimed at the 14- to 17-year age group. Although a quarter of the book focused on what a nice man Lenin was and how lucky the Soviet Union was to have Communism, the rest contained concrete instruction on a variety of subjects: military structure and recognition of vehicles and aircraft, rank and insignia recognition, and weapons handling and marksmanship covering the following weapons:

Type	Caliber	Designation
TOZ-8	.22 cal.	Bolt-action rifle
TOZ-12	.22 cal.	Bolt-action rifle
AKM/AKMS	7.62mm	Assault rifle
PPD-40	7.65mm	Submachine gun
PPSh-41	7.65mm	Submachine gun
PPS-43	7.65mm	Submachine gun

Other subjects covered in the manual were first aid, reconnaissance and intelligence gathering, semaphore, morse code, construction of simple transceivers, operation and maintenance of the military transceivers R-105M, R-108M, R-109M, and the

TAI-43 field telephone. Patrolling formations, fieldcraft, and basic survival skills—including navigation by compass, sun, and stars—are covered. Civil defense skills covering traffic management and fighting fires with syringe pumps, hoses, and OVP-5 and OP-5 fire extinguishers are covered in depth. Nuclear, biological, and chemical defense are also covered, along with the use and maintenance of the GP-5, DP-6, DP-6m, and R-2 respirator masks. Badge award standards for athletics, motor cross, motorcycle cross, fixed-wing and rotory aircraft pilot's license, parachuting, and scuba diving also receive coverage. The Boy Scouts would be hard pressed to match the variety of skills training available.

Appropriately enough, the CPSU has its own premilitary training that also leads to adult membership in the CPSU—which, as I have said, is a prerequisite for anyone working in special assignment units because it implies political correctness and reliability. Children from the age of seven can step onto the first rung of CPSU membership by joining the Octobrists; at the age of 10 they can move to the Pioneers, where they participate in drills, marches, and guarding war memorials, as well as learning tactics, civil defense, first aid, military discipline and regulations. During their annual *Zarnitsa*[14] war games, they get to handle unloaded weapons and learn about military vehicles. "A Pioneer reveres the memory of fallen fighters and prepares to become a defender of the motherland," the youngsters are constantly reminded.

At 15, they can join VLKSM,[15] known as Komsomol. Young men remain in the movement during their conscript service. At 29 years they can join the CPSU. They participate in annual *Orlyonok*[16] war games and receive training in weapons handling; radio communications (including finding of covert transmitters by triangulation); grenade throwing; shooting; and company-, platoon-, and squad-size operations.

Every year from January through March, all young men who have reached 17 are required to register for conscription at their local military commissariat or *voenkomat*. The job of the *voenkomat* is to gather files on the young men after contacting their schools, the MVD, KGB, DOSAAF, and the Komsomol. The file contains information on their educations achievements, leadership skills, family backgrounds, ethnic origins, political backgrounds, special skills, and career plans.

The Special committees (Spetskomy) of the GRU, MVD Interior Forces, and KGB Forces use the file as a means of preliminary selection. Their selection standards are as follows:

- capable but uncomplicated young men, mostly from the farms or small towns of Russia rather then the ethnic minorities
- preferably DOSAAF volunteers with parachute training
- generally about 10 years of secondary education (intellectuals are not normally welcome but healthy, resourceful, fast learners are)
- weight between 130 and 180 pounds, with a minimum height of 5 feet 7 inches

When conscripts report to their center for conscript service, they are not informed of initial selection for special assignment duties. When they reach the unit for which they have been designated, then the political officer or *zampolit* will inform all the conscripts. This does not guarantee placement in the unit.

Regardless of whether they enter special or conventional units, the conscripts then enter two programs. The first is a two-to-three-week initial training program called the "young soldiers course." At some time, they enter the VSK[17] program, first introduced in 1965 but later upgraded in 1973. The program is designed to prepare soldiers for the rigors of combat and to encourage them to take up sports in their free time. To this end, a system of awards and badges exist. However, physical fitness testing takes place several times during the conscripts' service. Servicemen are required to participate daily in physical training during the course of their six-day working week. In addition they must:

a. have a theoretical knowledge of a number of physical fitness programs operating in the USSR

b. know how to explain and perform a minimum of two routines from the *USSR Physical Training Manual*

c. participate in at least five Olympic sports competitions

d. participate in the pentathlon and and special-unit biathlons

Special assignment troops must be able to cover up to 30 miles a day with heavy operational loads regardless of terrain. They must also be able to swim at least 160 feet with 60 pounds of equipment. Current schooling requires expert knowledge and handing of the following Soviet weapons systems:

Type	Model	Caliber
Pistol, semiauto	PM	9mm
Pistol, semiauto	PSM	5.45mm
Pistol, (silenced) semiauto	PM69	9mm
Pistol (silenced) semiauto and auto	APS	9mm
Carbine, semiauto and auto	AKS 74U	5.45mm
Assault rifle, semiauto and& auto	AKS 74	5.5mm
Sniper rifle	SVD	7.62mm
LAW	RPG 7D	40mm
LAW	RPG 16D	58.3mm
LAW	RPG 22	72mm
LAW	RPG 26	72mm
Grenade launcher (under AKS 74)	BG 15	40mm
Grenade launcher (belt-fed)	AGS 17	30mm

Military transceivers consist of the R255PP personal radio and the R354M manpacked unit radio. Every battalion or *otryad* has its own signals company equipped with R360, R361, R357, and R358 communications equipment or the R148 communications vehicle.

Within initial training, about 25 percent of conscripts will be selected for training as noncommissioned officers, and of the remainder about 20 percent will be rejected by the training unit staff.

Close-quarter-battle (CQB) training plays a very large part in the training of special assignment soldiers. Generically, it is called *rukopashnyi boi* and has two main components: unarmed combat and skill at arms. The skill-at-arms aspect covers firearms use at close quarters, including a separate section called "cold weapons."[18] The latter teaches the use of the knife (*nozh*) and concerns two types: NRS 1 and NRS 2.

NRS 1 (knife, reconnaissance, special, type 1) fits into a

sleeve in the right thigh pocket of trousers and is issued to special units. NRS 2 can also fire a 7.62mm pistol round to an effective distance of 65 feet, in addition to functioning as a knife. Troopers are then schooled in the use of the entrenching tool (*shantsevaya lopata*) as a weapon and the bayonet (*shtyk*). The men are also taught to throw all these weapons accurately.

A particular favorite for throwing is a small sharpened steel plate 15-centimeters square for throwing distances of 20 to 75 meters and is simply referred to as a "plate" (*plastina*). This first made its appearance in Naval Infantry manuals in 1982, and then in 1986 the Military Institute of Physical Culture in Leningrad published it in the series, *Special Features of Physical Training of Naval Personnel (Part 2)*.

Unarmed combat is based on a mixture of the Soviet combat sport of *sambo*[19] mixed with judo and karate. The karate is very similar to Korean styles, which is unsurprising because North Korean instructors were responsible for its introduction into Russia. The KGB started karate training in 1975 as part of obligatory and basic training. Every year the airborne forces sponsor a championship in their form of full contact, which the special assignment units can enter. This tends to be a no-holds-barred competition with minimum protection, and striking continues even when one is on the ground.

During visits to the Higher Airborne Forces (VDV) College at Ryazan in the Moscow Military District, I was able to participate in *rukopashnyi boi* training under instruction from Lt. Col. Vladimir Panteleev, the college's chief instructor and former VDV CQB champion. This included the protocols of throwing "cold weapons." Ryazan has been wrongly identified by the Western media as a Spetsnaz center. Certainly its language center provides instruction for Spetsnaz, and it shares training facilities at a nearby training area with the Sixteenth Spetsnaz Brigade based at Chuchkovo. The school trains officers for the airborne forces and usually has 2,000 *kursanti*[20] spread over its four-year course.

Applications for the 500 first-year places each year normally exceed 32,500. The applicants have to be between 17 and 21 years (or 25 if they have previous military service). During the course of the year the Spetsnaz Committee of the Third Department of the Fifth Directorate of the GRU visit the college to headhunt potential Spetsnaz officers, as they do the reconnaissance faculty of the Suvorov and Frunze military colleges in Kiev, the Kiev Military College, and the Military Institute of Foreign Languages in Moscow. Other officers will be headhunted after graduations and placement with their units. GRU Spetsnaz maintains two NCO training brigades and one college for training officers.

SPETSNAZ

GRU Spetsnaz officers were trained at the KGB's Special Tasks School at Balashikha near Moscow until the early 1970s, when a school was formed at Krasnodar in the North Caucasus Military District. It was named after General of the Army Sergei M. Shtemenko, postwar pioneer of the GRU Spetsnaz.

A normal Spetsnaz brigade consists of between 400 and 1300 men divided into 200-man *otryady*. Each *otriad* contains three companies plus a signals company. Each company has three groups; each group finally subdivides into three patrols of four to five men, each called an *otdelenie*. In time of war, each brigade is brought up to full strength by recalling reserves from civilian life. Normally, independent Spetsnaz companies attached in support of conventional units consist of 140 personnel, 111 of which are conscripts. The ratio of officers to men is usually twice as high in special assignment units, usually one officer for every 12 men as opposed to one officer for every 25 men in conventional units.

The defense forces like to maintain that their job is solely concerned with the defense of the USSR and that they leave the messy job of internal security to the MVD and KGB, but this is not supported by fact. The very origin of army Spetsnaz is in internal security work.

In 1927 a 15-man special diversionary unit parachuted into the Saksaul of Kazakhstan to operate against Muslim separatists. Further parachute operations followed in 1929 and 1931. Airborne forces came into being in August 1930 with the role of diversionary reconnaissance. In February 1932, a document regulating their role listed ambushes and behind-the-lines sabotage of enemy headquarters and logistics as their primary roles. The first name given to airborne forces (now the VDV) was *Brigady Desantnykh Spetsial'nogo Naznacheniya* or Airborne Assault Special Assignment Brigades. By 1938, the USSR had five airborne corps, and each corps had one to two special mission battalions.

In November 1936, the first Spetsnaz detachment staffed by KGB and GRU personnel was started in Spain during that country's civil war. Besides Soviets and Spanish Communists, about 100 foreigners from the International Brigade were directly recruited. In 1937 all KGB/GRU special operations units were incorporated within a new Fourteenth Special Corps, which was disbanded after the end of the Spanish Civil War, with some 300 Spanish members joining special assignment units in the NKVD of the Soviet Union. A GRU veteran of the Spanish Civil War led a 50-man Spetsnaz unit against the Finns during the Winter War of 1939-1940, but with little tangible success.

Then came World War II, with special assignment units deployed by NKVD behind both German lines and their own, and by the GRU through both the army and Naval Infantry. After the war, GRU Spetsnaz was disbanded, and all special assignment tasks fell to the NKVD or the MVD, as has been described. GRU Spetsnaz was revitalized in the late 1950s at about the same time the NATO nations were busy developing their special operations capabilities after a postwar lack of interest in the special forces field. Hence, the need arose to train GRU Spetsnaz officers for a while in the KGB special tasks school near Moscow. The GRU Spetsnaz were not in position in the scheme of things to be involved with the suppression of rebellion in the Eastern bloc immediately after 1945. From 1945 the NKVD operated special assignment units throughout Estonia, Latvia, Lithuania, eastern Poland, Ruthenia, Bukovina, and Bessarabia, territories recently annexed to the west of the USSR, as well as already annexed areas such as the Ukraine.

These antipartisan campaigns were fought practically into the 1960s. In the Baltic States, the U.S. Air Force flew secret missions supplying the guerrillas, but unfortunately the CIA shared its operations with MI6, which was riddled with Soviet spies and sympathizers. When East Germany rose in rebellion in 1953, the KGB and MVD were on hand to put the rebellion down. KGB and MVD special units ensured the annexation of Poland and Czechoslovakia. In March 1948 Czechoslovakian Foreign Minister Jan Masaryk, who advocated independence for his country, was found dead in the courtyard of his residence. The USSR claimed he committed suicide; the Czechs believe a KGB special unit was responsible. When Polish leader Wladyslaw Gomulka proved too independent, he was replaced. In 1956 when Hungary rose in rebellion, there was no GRU Spetsnaz on hand to act, so the head of the KGB, General Ivan Serov, launched a campaign to seize Hungarian Prime Minister Imre Nagy and his advisors. Serov arranged a Soviet-Hungarian official dinner in Budapest that Nagy and his people attended. Halfway through the dinner, Serov and his Spets group from the Balashikha school and the Department of the 1st Directorate —which had inherited the Central Committee special tasks—did the dirty deed. (A similar tactic was used in December 1979 to seize Afghan government officials during the Soviet invasion.)

Serov was a Frunze graduate and member of the Soviet army. In 1939 he transferred to NKVD and served under General Kobulov, the organization's counterinsurgency specialist. Kobulov's NKVD *istrebitel'nye otryady* supplied the security at the Yalta conference, protecting British Prime Minister Winston Churchill, U.S. President Franklin D. Roosevelt, and Soviet leader Generalissimo Joseph Stalin. Stalin and Churchill then went straight to the Ukraine and Byelorussia on guerrilla-extermination duties. Serov was sent to the Baltic States to direct similar duties there. Later he transferred to the Ukraine where he befriended Nikita Khrushchev, who was to promote him to chairman of the KGB.

By 1968, however, the GRU Spetsnaz was in full operation and was used in the seizure of Prague, working alongside the KGB's Department V, which had replaced Department 13, and elements of the MVD Spetsnazovtsy on loan from the Felix

PREFACE

Dzerzhinsky Division. GRU Spetsnaz from the brigades based at Kirovograd and Mariinogorko were earmarked to work with KGB Department V personnel and personnel from the Balashikha school. They would be supported by the 103rd Guards Airborne Division based at Vitebsk. Soviet logistics specialists depleted Czechoslovakia's military supplies through contrived Warsaw Pact exercises in that country, East Germany, and western USSR. The Warsaw Pact then scheduled exercises in Bohemia to divert the Czechoslovak military from intended crossing points. Friday 16 August 1968, all special assignment personnel were placed on standby for the invasion and started discreet preparations. On 20 August 1968, Aeroflot aircraft began landing at Prague's Ruzyne airport. Just before 8 P.M. an AN-24 from Moscow arrived containing communications and signals personnel and the Spetsnaz headquarters element. At 9:30 P.M. another unscheduled AN24 from Lvov containing the Department V personnel arrived.

They were met by Colonel Elias of the Czechoslovak Interior Ministry and Lieutenant Colonel Stachovsky of the Border Guard, representing the Czechoslovak counterparts of MVD and KGB. The KGB Osnaz departed for the Soviet embassy in Prague and its AN24 returned to Lvov. At 12:00 P.M. the Czech Interior Ministry personnel closed the airport, and the KGB Osnaz arrived with Czech-speaking Soviet officers. The Osnaz, though dressed as civilians, openly carried weapons and seized the control tower, foreign departures, customs, and airport communications. Two Aeroflot AN-12s carrying the GRU Spetsnaz landed and taxied to the administration building. The Spetsnaz linked up with the Osnaz troops and swept through the airport, driving all airport personnel and tourists out of the building. Women and children were then allowed to return. At 5:30 A.M. on 21 August, the Spetsnaz group allowed airport personnel and tourists to leave.

While this was happening, three GRU Spetsnaz from Mariinogorko Brigade arrived with the lead elements of 103rd Guards Airborne Division. One of the first VDV units to arrive was 271st Airborne Artillery Regiment, which moved to a position overlooking Prague and positioned its artillery. The rest of the 103rd Division—the 393rd, 583rd, and 688th Guards—occupied key points such as the brigades over the Vltava River, railway stations, post office, central telephone exchange, and central crossroads. The MVD Spetsnazovtsy and GRU Spetsnaz captured the television and radio stations. GRU Spetsnaz accompanied by Department V Osnaz seized the presidential palace. At 3:00 A.M., GRU Spetsnaz seized Prime Minister Oldrich Cernik at bayonet point in the government presidium building. A security force from the 103rd Guards arrived and secured an outer perimeter.

At 4:00 A.M. three BTR-50s led by a black Volga sedan left the Soviet Embassy and went to the Czechoslovak Communist Party Central Committee building. A security force from the 103rd Guards Airborne Division arrived and secured an outer perimeter. GRU Spetsnaz leaped out of the BTRs and were led into the building by two KGB Osnaz and Col. Bohumil Molnar of the Czechoslovak Interior Ministry. President Dubcek and his advisors Smorkovsky, Kriegel, Spacek, Simon, and Mlynar were meeting in Dubcek's office when the door was kicked in and eight Spetsnaz rushed in and placed their weapons at the back of the politicians' heads. The KGB men then walked in with the Czechoslovak colonel and announced that they were all under arrest. They were then flown under Spetsnaz and Osnaz escort to Moscow.

The invasion of Czechoslovakia was the first major operation in which KGB, MVD, and GRU special assignment units had worked together. Department V and its predecessors, however, had been very active prior to this. One example was the attempted assassination of the shah of Iran in 1962. This was masterminded by the then chief of Department V, Lt. Gen. Ivan Fadekin. (Fadekin had been a deputy commander of partisans in the NKVD during World War II.) Following the CIA-instigated coup in Iran the Central Committee of the USSR decided to eliminate the shah. Fadekin was given the task and arrived in Teheran in 1961. In February 1962, he arranged for a Volkswagen packed with explosives to be positioned along the route of the shah's convoy between his palace and the parliament building. The explosives had a radio-initiated detonator. As the shah's car passed, the radio signal was sent but failed to detonate the explosives.

The primary roles of Department V were assassination, sabotage, and subversion in time of war and linking up with GRU Spetsnaz groups. After the defection of a Department V officer to MI6 in 1971, the department was closed down. Its direct-action role passed to a newly created Eighth Department within the S Directorate of the 1st Chief Directorate formed in 1973. However, its role, codenamed "Kaskad," was primarily limited to wartime as the Cold War thawed on both sides of the pond.

Prior to the Soviet invasion of Afghanistan, the Eighth Department was ordered to kill President Amin of Afghanistan. In September 1979 Lt. Col. Mikhail Talybov of the 8th Department was provided with Afghan documentation and sent to do the job. He managed to get a position as a cook and tried to poison Amin on numerous occasions. His failure led to an invasion similar to that of Czechoslovakia in 1968.

In 1977, with the approach of the 1980 Summer Olympic Games in Moscow, the MVD decided it required an antiterrorist intervention capability on par with the German GSG 9, French GIGN, British 22 SAS CRW (Special Air Services Counter Revolutionary Warfare) teams, or U.S. SWAT teams. Far from being the model of law and order, the USSR had the world's worst record for air piracy, now called skyjacking. The Interior Ministry turned to its elite Felix Dzerzhinsky Motor Rifle Division based near Moscow to come up with a solution. On 31 December 1977 the Special Assignment Company[21] of the division was formed under its commander, Captain Mal'tsev.

Recruits came from within the division, and only 50 out of 100 passed the selection course. Mal'tsev chose maroon berets, badges, shoulder boards, and hat ribbons since that was the color of the Internal Forces. The berets are solid maroon or mottled as the Russians prefer. The unit became known as the "mottled berets."[22]

(MVD Internal Forces special assignment personnel are not referred to as Spetsnaz; they are called Spetsnazovtsy, according to Col. Gen. Yuri Shatalin, who in 1990 was head of all MVD Internal Forces. On 19 April 1990 he listed the number of special assignment troops at his disposal at 2,500, but this might include the separate OMON units that I will discuss later.)

In 1982 the unit carried out its first internal operation. It flew to the city of Sarapul in the Russian Republic and from there flew 70 kilometers northwest to Izhevsk, where they transferred to buses and drove to a school where two criminals were holding children hostage and demanding to be given an aircraft to fly to the West. Working with the local KGB, they negotiated the surrender of the criminals and release of the hostages.

Then in 1987 Interior Ministry troops from the Siberian city of Perm seized an aircraft, attempting to fly to the West. Sgt. Nikolai Matsnev, the gang leader, persuaded five other conscripts to join him, two from his base and three from another. Three of the plotters were drug abusers, using both opium and hashish. Matsnev and Privates Konoval and Yagmurzhi took weapons and ammunition from their armory and set off on foot for another camp where the other three were stealing an armored vehicle. En route, an MVD militia patrol car stopped them, and they killed the two sergeants inside. Konoval fled and was later arrested. Matznev and Yagmurzhi took a taxi to the airport and shot their way through security to the aircraft, a TU-134 destined for the Siberian oil fields. In the process they damaged the fuselage and killed two of the 74 passengers. They demanded that the pilot fly them to the West, and he agreed to do so once the aircraft was repaired. Matsnev released the women and children and other hostages. Yagmurzhi demanded opium and a guitar from the airport authorities and then promptly lapsed in to a drugged sleep. The Spetsrota then stormed the plane, killing Matsnev and wounding Yagmurzhi, but two passengers were also killed.

Among other intervention duties of the Spets group were security and rescue duties at Chernobyl when a nuclear reactor exploded in 1986.

In February 1988 Azerbaijani gangs swept through the town of Sumgait on the shores of the Caspian Sea, raping and killing Armenians and burning. Even Azerbaijanis who helped their Armenian neighbors were attacked and killed and their women raped. A story started that, on 27 February, two Azeris had been killed by Armenians. A riot started, and between 60 and 100 people were killed. The Azerbaijani Interior Ministry moved slowly to quell the rioting, so the 104th Guards Airborne Division was

Soviet Spetsnaz learned how to use their entrenching tools as weapons in the KGB Alpha Team Manual. Photo courtesy of Jim Shortt

sent in to crush the rioting and the MVD Spetsnazovtsy went in to rescue Armenians in the town.

On 9 April 1989, they were in Tbilisi with the members of the 104th Guards Airborne Division from Kirovabad to quell nationalist unrest. The local Interior Ministry troops were sent in under the command of Interior Forces General Yuri Yefimov, who asked the Defense Forces District Commander Colonel General Igor Rodionov for assistance. Twenty civilians died from entrenching tool injuries and toxic gas. The Defense Forces admitted using 27 canisters of a riot-control agent called *cheryomukha* but denied that it was toxic, comparing it with tear gas. Statements taken from one member of the Spetsrota present were published in a Lithuanian newspaper on 4 May. He spoke of members of the unit putting on their berets when they arrived and that there was only 100 of the special unit among the paratroopers and other MVD units.

The Spetsrota drew their entrenching tools and started to hack at the crowd regardless of sex or age. The soldier recalls it as a night filled with terror he will never forget. Soon after, on 24 April 1989, the unit was sent to Perm on the Kama River in the Urals. Three zeks, prison camp inmates, had taken an MVD captain and three female staff hostage.

They were armed with prison picks and were demanding weapons, ballistic vests, and safe passage out of the USSR. A KGB negotiation team tried fruitlessly to get them to surrender. Misha Komisarov, a member of the Spetsrota, describes what happened: "That morning we were told to fly to Perm and arrived at 19:00 hours when it was beginning to get dark. The zeks said they would start killing the hostages in 90 minutes. There were 30 of us in the assault group[23]; we prepared diversionary munitions[24] and cutting charges.

"When zeks killed the captain, we stormed the building, wearing body armor and titanium helmets for protection. We cut an entry port in the wall with explosives. The first unit went in and seized two prisoners. Down the corridor other zeks held the women. We knocked the door down, threw in stun grenades,[25] and released the women hostages."

MVD INTERNAL FORCES

Thus started a summer of pogroms that took the MVD Spetsnazovtsy around the USSR in support of MVD internal troops. The first was in Uzbekistan in May 1989, where eventually 12,000 MVD Interior Forces had to be deployed. The unit was deployed to Fergana and supported by the Independent Airborne Regiment formed from the old 105th Guards Airborne Division. This regiment consisted of veterans of the Afghanistan War and

PREFACE

Captured KGB Alpha Team titanium helmet and stun grenade. Photo courtesy of Jim Shortt

specialized in mountain warfare reconnaissance. In June 1989, they moved on to Novyi Uzan in Kazakhztan for more ethnic firefighting. The next month, they were sent to Abkhazia in southern Georgia to rescue hundreds of Turk-Meshkhetis people from Georgian fanatics who wanted to murder these Muslims. Legend has it that 10 of the men from the unit defeated 100 rioters in hand-to-hand fighting.

NATO has always estimated that the MVD Internal Forces numbered some 30 divisions, or about a quarter of a million troops in July 1989, at the height of ethnic unrest in the USSR. Interior Minister Vadim Bakatin stated that he had a resource of 700,000 militia and only 35,000 troops, of which only about

18,000 poorly equipped men were available for rapid deployment—including a now airmobile unit formed within the Felix Dzerzhinsky Division. However, in 1990 this was contradicted by the commander of the Internal Forces, Col. Gen. Yuri Shatalin, who revealed that the overall strength of MVD Internal Forces was 350,000.

A breakdown of units showed that 15,400 troops had been sent to the Transcaucasian republics to deal with ethnic unrest. This rose to 25,000 by the spring of 1990. Added to this, the 104th Guards Airborne Division was transferred to the control of the KGB Chief Directorate of Border Guards and operated in the disputed enclave of Nagorno-Karabakh, where at least 10,000 MVD troops were then based, including the now airmobile unit. On 19 April 1990, General Shatalin said that the MVD had Spetsnazovtsy assets of 2,500 men, but this may include the MVD militia units called OMON, which strictly were not part of the Interior Forces.

However, Interior Forces in the various republics did set up their own maroon-bereted units. I have come across these in Latvia. In March 1992, I had returned to Tallinn and Riga to teach an explosives course using Semtex and Soviet TNT. At the close of the course, I met with the Latvian prime minister before the rest of the team and I were taken to a farewell dinner, during which members of the group were attacked by a larger number of Russian mafia.[26]

The fighting was with whatever weapon came to hand, but the area was too confined to draw firearms. While I was trying my soccer technique on someone's scrotum, a gangster tried to get behind me with a knife, which I presume he wanted to freely give to my back. However, he had forgotten the Sixteenth Law of Murphy: *Never take a knife to a gunfight.* Andy Karn, one of the IBA Team, drew his two-inch barrel Smith & Wesson .38 revolver and shot him. The ball went through his left chest and through the hand of another villain who was pushing the thug forward, through the cheek of the gang boss, and then broke a mirror. Actually, the first gangster was rather lucky, as the next round had been part of a cocktail mix and was a hydroshock. The gangster broke and fled, but within 90 seconds a maroon-bereted Interior Ministry unit arrived in body armor with AKS-74s and AKS-74Us and war dogs. *These* Russian troops had been paid to cover the gangsters' retreat if things went wrong. What the mafia hadn't known was that they were pitched against government security forces, not tourists.

There is now an investigation into the activities of the unit. Preliminary findings indicate that the mafia thugs were from the Interior Ministry, they wore maroon berets and camouflage, and they were festooned with weapons. They were not, however, from the Felix Dzerzhinsky Division. Members of the latter unit make reference to imitators and how they give their unit a bad name. They recall how in Semipalatinsk in Kazakhstan a unit was sent to rescue hostages, resulting in the death of the unit's major and some hostages and the wounding of the second in command (a captain).

OMON

In 1987, the first OMON[27] unit appeared in Leningrad with the support of the city's CPSU first secretary. They were formed from the Militia Patrol Regiment and were seen as a protector of the CPSU at a time when Communism was becoming less and less popular within the USSR. The Leningrad OMON adopted black berets as their headgear since there was a plentiful supply through the local Naval Infantry, and a number of the OMON had served in it. When the first secretary, Lev Zaikov, was promoted and moved to Moscow, he suggested that the OMON be set up in the capital. Initially formed from the Moscow 2nd Police Patrol Regiment, the new Moscow OMON chose gray berets to match the gray militia uniform.

OMSN

Another special unit, the OMSN, a plainclothes antimafia group within the MVD, was already in Moscow at 38 Petrovka Street. Often films and photos of this highly professional unit

OMON beret and insignia with black beret (chornyi beret). Photo courtesy of Jim Shortt

wrongly bear the caption of OMON or MVD VV. The Soviet Ministry of the Interior regularized the formation and operation of OMON as police special assignment units similar to S.W.A.T. and antiriot teams under Regulation 4603 in August 1988. While Interior Forces special units were centrally controlled, this gave authority to the ministers of the interior of the 16 republics that made up the Soviet Union. Some republics like Estonia told the ministry that they had adequate police resources to deal with their problems and did not form OMON units. Others like Lithuania, Latvia, Moldavia, and Azerbaijan formed units. OMON units were set at between 100 and 300 men equipped with AKS-74Us, Makarov pistols, ballistic vests, riot shields, batons, and, normally, VDV camouflage uniforms. OMON troops were first used on 21 August 1988 against prodemocracy protestors in Pushkin Square. From the beginning, OMON units were to be under the control of the various republics' MVD and not the USSR.

In March 1988, the Ovechkin family of eleven smuggled weapons on board an Aeroflot TU-154 in Irkutsk and hijacked it in an attempt to get to London. They touched down in Leningrad, but the militia pretended they were in Helsinki. After two hours of negotiations, a five-man OMON team stormed the aircraft. The hijackers detonated an explosive device in the rear of the plane, turning it into a fireball and killing nine people and injuring 19.

In October 1988, Vladimir Kryuchkov became chairman of the KGB and the Kaskad program of the Eighth Department. He immediately formed a new KGB intervention unit based in KGB HQ at Dzerzhinsky Square in Moscow. The CO was Col. R. Ishmiyarov, and the assault term commander was Maj. O.G. Aliyev. The team was provided with equipment similar to that used by the GRU Spetnazovets, plus a variety of silent weapons including the AP5 Stetchkin 9mm, which converts to full automatic. Volunteers came from the KGB Border Guard Service and received training at the Balashikha school.

The team was trained in *rukopashnyi boi* and related physical methods by A.I. Dolmatov of the Central Dynamo sports club in Moscow, who produced the manual you are about to read. For this manual, he borrowed in part from the two 1945 GRU Spetsnaz texts for reconnaissance and prisoner handling, and reconnaissance in mountains.[28] From the start, it was obvious that the role of the Alpha teams and their MVD counterparts was not confined to humanitarian intervention. They also had a war role linked to Kaskad, which was run by Lt. Col. I. Morozova at this time. That role covered the traditional NKVD functions of fighting behind enemy lines and dealing with dissidents and enemy special forces.

MVD Spetsnazovets of the Spetsrota of the Dzherinsky Division in physical training. Note the Dynamo T-shirts. Photo courtesy of Novosti Information Agency

ALPHA TEAMS

Alpha teams first came to public attention on 1 December 1988, when four criminal seized a bus containing 30 school children in Ordzhonikdze and ransomed them for an IL-76T aircraft, money, and weapons. The operation ended on 2 December 1988 at Ben Gurion Airport in Tel Aviv when the Israelis convinced the criminals to surrender. The Israelis handed the hijackers over to the KGB Alpha team, who flew them back to Moscow. On 30 March 1989, 22-year-old Stanislav Skok hijacked a domestic

PREFACE

Moscow OMON units training in rappelling, entering buildings, and rukopashni boi. Photo courtesy of Novosti Information Agency

KGB ALPHA TEAM TRAINING MANUAL

PREFACE

flight from Astrakhan to Baku with 76 passengers and crew. He claimed to have a bomb in his bag and ordered the aircraft to land at Bina Airport in Baku, demanding a ransom and flight to Pakistan. At 11:30 A.M. on 31 March 1989, Major Aliyev and four other Alpha team members wearing Aeroflot coveralls entered the front of the aircraft. When the hijacker stepped away from the bag to light a cigarette offered by the team, he was seized and subdued.

AFGHANISTAN: A TRAINING GROUND

Afghanistan was the catalyst for all special assignment units, whether GRU, KGB, or MVD. After the Eighth Department's Lieutenant Colonel Talybov failed during 1979 to kill President Amin, the USSR decided to invade and take over Kabul. Since the May 1979 murder of Soviet advisors and their families in Herat, two Soviet special divisions composed of Uzbek and Tajik nationals and dressed in DRA (Afghan army) uniforms positioned themselves along the Soviet-Afghan border. These two divisions were chaperoned by the 105th Guards Airborne Division and reinforced by two regiments borrowed from l03rd and 104th Guards Airborne Divisions.

The invasion of Afghanistan started on 28 December 1979. At this stage, the Eighth Department was depleted and had to borrow KGB officers from other departments of the First Chief Directorate. They also borrowed personnel from the Felix Dzerzhinsky Division Spetsrota and the Spetsnaz Brigade based at Chirchik, along with the lst Battalion of the 16th Spetsnaz Brigade at Chuchkovo. The KGB team was led by Colonel Boyarinov, the Balashikha school commander, while the overall pacification of the Afghan administration was under the command of Maj. Gen. Viktor Paputin of the MVD.

As in Czechoslovakia, Afghan notables were invited to a reception at the Soviet Embassy in Kabul, where they were seized and locked in the cellar by a Spets group. An airborne battalion provided the security group and perimeter for the Spets group, which landed at Bagram Air Base. Soviet advisors had disarmed most Afghan military units under the guise of an equipment check, as had been done in Czechoslovakia. The group using the Spetsnaz favored BTR APCs escorted by VDV BMD-1s and headed for the Darulman Palace. An Afghan sentry who opposed them was dispatched with a silent Stechkin pistol.

On reaching the palace, the BMDs rammed the palace gates, where one stalled and stuck. The paras secured the perimeter, eliminating the palace guard. The KGB/MVD/GRU Spets group led by Colonel Boyarinov stormed the palace and ran into heavy opposition and started taking casualties. Boyarinov ran out to get the paras to bolster the assault, forgetting his order for them to kill anyone who ran out of the palace. So the VDV killed the Spets mission commander. The assault continued, and everyone in the palace was killed by the group except Talbov, the Eighth Department assassin who hid under the stairs until, in his own words, the voices "sounded human again." Other Spets groups seized the radio and TV stations and vital government centers.

The primary principle here is that groups such as the Spetsrota that are trained to enter buildings, kill terrorists and criminals, and rescue hostages are more capable in entering building and killing people when they don't have to worry about hostages.

All three services provided special assignment groups while the war lasted. The KGB trained assassins and saboteurs and fielded special border guard units to intercept and ambush mujahideen supply caravans. The MVD-trained Afghan Interior Ministry special assignment troops and the GRU-supplied Spetsnaz brigades intercepted supply caravans. The Russian nickname for the Spets groups in Afghanistan was *okhotniki karavana*, caravan hunters. Periodically, they raided mujahideen concentrations and bases, and toward the close of the war they formed special "Stinger" hunter teams that were airmobile. They were heavily involved in Operation Mistral, the Soviet clearing of the road from Gardez-Khost. I was in Paktia province serving with National Islamic Front of Afghanistan (NIFA) mujahideen at the close of that operation, and certainly the greatest threat apart from being shelled and bombed came from Spetsnaz ambush.

Since returning from fighting the Soviets in Afghanistan, I have been to Ryazan' to undergo their training, drink vodka

Jim Shortt in the Gardez-Urgun-Khost area during the war in Afghanistan. Photo courtesy of Jim Shortt

and *samogonka* (moonshine) with them, and share their *banya* (steam bath). Since International Bodyguard Association (IBA) first became involved in assisting the Baltic States in 1989, I have had many former GRU and MVD Spetsnaz and Spetsnazovtsy under my command, and several have become dear friends. Many are veterans of the Afghan war and count me as an *afghanets*[29]—even though I fought against them. I can say that that war left a mark on the soul of anyone involved, just as Vietnam did on Vietnamese and Americans alike. The history of special operations in Afghanistan has yet to be written, but for the purposes of this book, this quote from GRU veteran Lt. Col. Sergei Balyenko of the Spetsnaz staff of the Fifth Directorate sums up how they saw it: "Afghanistan was a polygon—a training ground."

During and following the Afghanistan War, MVD and KGB special assignment units received a large influx of former Spetsnaz and paratroopers, according to the chief of Internal Army troops, General Shatalin. When creating the OMON units, many commanders selected only those with special assignment unit combat records. However, Moscow OMON, which boasted of being packed with *afghantsy*, had mostly former militia men. Its commander, Colonel Ivanov, was nearing retirement age in 1991, and the unit contained men well over 40 even though the official line was that OMAN consisted of fit, young paratroopers between the ages of 23 and 26.

Other problems troubled the OMON forces as well. Firearms training was limited to 30 rounds per man per year by order of the Interior Ministry. On 2 August 1990, former Spetsnaz participating in Airborne Forces Day beat up 15 OMON troopers sent to arrest them and then went on a wrecking spree while the Interior Ministry held the 150-man OMON reserve unit back. This incident caused a spate of resignations from the Moscow gray-beret unit.

Pay in 1991 was 260 rubles a month, about $2.60 for an OMON recruit, yet in 1991 Moscow OMON numbered 1,500 men.

In the Soviet press of 30 May 1991, prominent Soviet figure Dr. Alexei Kiva warned that the conservative military-backed Soyuz group in Parliament was "a mixed bag of dictators" and were planning a coup d'état. His article is a precise analysis of the forces at work in the USSR—a collaboration of conservative Communists who had lost power and influence under Gorbachev and aligned themselves with a new breed of Russians who wished a return to the previous glory of the old Russian Empire devoid of Communism. These Russian imperialists viewed the Baltic States, the Caucasus, Poland, and even Finland as part of the historical Russian empire, with the same fervor as the Islamic fundamentalists speak of reestablishing the Islamic Crescent (Spain, North Africa, the Middle East, Malaysia, Indonesia, and the Philippines) before the final battle with the "Christian" nations.

The Soyuz group received widespread support from within the Soviet military, whose professional officer class had been especially hard hit. Their budgets were radically cut, and Gorbachev had initiated troop withdrawals from East Germany. Poland, Hungary, and Czechoslovakia, where living was good for Soviet officers and their families. Back in the USSR, their wages were slashed, and no housing was provided for their families. Invariably, officers had to live in an officer's hotel[30] while their families went to live with their parents, which could be on the other side of the USSR.

In a rebuttal to Kiva, the cochairman of Soyuz, Colonel Viktor Alksnis, accused the democrats of being the real coup threat. Alksnis, a member of the Latvian and Soviet parliaments, was a military officer based in the Baltic States. The events that led to the August 1991 coup attempt were centered on the Baltic States and are far more sinister than the public has been led to believe.

The impetus for the coup came from three major sources: the defense minister, KGB chairman, and interior minister[31]. These three controlled not only the conventional forces but also the special assignment forces of the USSR. During the attempted coup, they used all these units in the fight against democracy in an illegal manner—even by Soviet law. Not only did their actions overstep the law, they also used forces that clearly were outside legal boundaries, the Russian mafia.

RUSSIAN MAFIA

The Russian mafia is no relation to the Cosa Nostra or Italian mafia. In Russian, "mafia" is a generic term for racketeers, black marketeers, and gangsters. The mafia organizations are becoming even more organized, and although the current situation is more akin to Chicago in the 1920s with gang fights over territory, it will develop into something bigger, more violent, and more sinister than its Italian counterpart.

The Russian mafia was opposed to the breakup of the Soviet Union because that reduced the mafia's power. Mafia leaders feared that corrupt nationalist officials would have more sympathy for gangsters from their own national groups than for "foreigners," such as Russians. Second, new national borders meant greater difficulty in transporting their illegal goods, thus more bribes would have to be paid. Under the existing system, if a Communist party boss in Estonia or Byelorussia wanted "hard currency" goods from the West, he got them in return for arranging favors for the mafia gangs. An Estonian nationalist politician might have more interest in his nation than himself.

The KGB and Communist party found the Russian mafia a useful ally. Soon after the Communist party in the Baltic States (Lithuania, Latvia, and Estonia) was voted out of power, the Russian mafia moved in to increase lawlessness, robbery, prostitution, and murder. The Communist bosses then turned around and said, "Look, under the nationalists, law and order breaks down."

The Soviets invaded the Baltic States much as they had Poland in 1939, which resulted in the Soviet-Nazi carving up of Poland. On the night of 14 June 1941, 60,000 men, women, and children were arrested in the three Baltic nations, herded into cattle cars, and shipped to Siberia. In the days and weeks that followed, thousands more were sent. After the invasion, the Soviets shipped in settlers from the other Soviet republics and did their best to eradicate the Baltic languages, music, and traditions. With *glasnost'*, the truth of what happened in and since 1941 started to come out.

Not until the late 1980s did the Estonians, Latvians, and

Lithuanians reemerge to retake control of their own destinies. On 30 March 1988, Estonia, though still occupied by the Soviet military, declared itmself a sovereign nation. Further acts toward independence followed, including restoration of the language. Then on 2 February, 1990 the Estonian parliament declared national independence. Similar laws were enacted in Latvia and Lithuania. The old-style Communists were voted out of power, and they linked up with their friends in Moscow and in the organs of government, the KGB, MVD, and the Defense Ministry. In the Baltic States they formed themselves into the Interfront and then tried to raise ethnic tensions between Balts and non-Balts by presenting their struggle as Balt against Russian rather than democracy versus communism.

The link between the men in Moscow (particularly Boris Pugo) and the direct action in the Baltic States was Colonel Nikolai Goncharenko, a leading Interfront activist and founder. In April 1991, Moscow appointed Goncharenko as the immediate superior of OMON units housed in Lithuania (Vilnius) and Latvia (Riga). Moscow had selected OMON to be the vanguard of the attack upon independence and prodemocracy organizations within the USSR.

The first deployment of the OMON was on 21 August 1988 against members of the Democratic Union in Moscow. On 2 November 1990, OMON based in the Moldavian Republic that borders Romania were used against independence demonstrators in the town of Dubossary, resulting in three civilian deaths caused by OMON fire. In a demonstration of how the conservatives in the Soviet administration were willing to exploit ethnic and national feuds, the Soviet 4th Army of the Caucasus assisted OMON[32] of the Azerbaijani Interior Ministry.

From 22 April 1991, they carried out attacks and deportations within the disputed Nagorno-Karabakh region, destroying Armenian villages. Soviet journalists who were present were told that the sweep was against Armenian Special Forces, which had hidden heavy weapons. This proved to be a lie because only hunting weapons belonging to a state farm and some police weapons taken from Armenian police officers were recovered.

PREFACE

On 6 May 1991, OMON and Soviet regular troops ambushed a busload of Armenian police, killing three, wounding three, and capturing six. The three wounded were later killed. The attack was a blatant cross-border violation. In the days that followed, further ambushes and village raids took place. The Interior Ministry official personally in charge at these operations was identified as Major General Andreev of the Central Criminal Department of the MVD. In late May, the USSR Procurator General's office declared the actions of the army and OMON to be criminal and started investigations into acts committed in the disputed border area. The activities of OMON units and their allies were not isolated and specific to that situation, but rather part of wider action in the Baltic States sponsored from Moscow and involving the Defense Ministry, KGB, and MVD.

Once the Baltic States had started on their road to freedom, the KGB stepped up its technical surveillance operations to gather intelligence of the Baltic States' intentions. They fed this to the Communist party in the Baltic States, the Interfront. Interfront had also formed its own paramilitary unit, which in Estonia was led by Mikhail Lysenko and paraded in paramilitary uniforms.

The OMON joined Soviet navy personnel in occupying the former Communist party building in the fashionable Riga suburb of Jurmala on 9 November 1990. They began an official but covert terror campaign that lasted until the defeat of the Moscow coup attempt in August 1991. Before this campaign, Communist hardliners had used their usual rent-a-crowd techniques to try and intimidate opponents and shout down those who disagreed. But these Interfront-organized tactics failed. Mobs packed with out-of-uniform soldiers, KGB, and mafia thugs found themselves outnumbered 10—and sometimes even 100—to 1 when they tried storming the nationalist parliament buildings. When they tried to sing Communist anthems, they were outsung with nationalist ones. Unarmed action hadn't worked, so they turned to armed action.

Following the action of the Riga OMON, the Latvian parliament stripped it of its police status in the republic. The USSR Ministry of the Interior immediately placed it under the orbit of the 42nd MVD Interior Forces Division based in the Baltic States, thus restoring its MVD powers. On 2 January, the Riga OMON stormed the Journalists' Union headquarters and raised a red flag over it. In Lithuania, OMON seized the old Communist party building in Vilnius. Its claim to legitimacy was that it was protecting CPSU property in line with President Gorbachev's decree.

In October 1989, while working in Scandinavia, I was briefed on the Baltic situation and asked whether IBA would be prepared to assist the Baltic governments in setting up security details. Up until 1988, the the KGB Ninth Department had supplied bodyguards, but the nationalist politicians obviously didn't want KGB minders and, therefore, had to form their own secret service details. They viewed assistance from Western government agencies as interference, but both the nationalists and Moscow would accept IBA.

IBA began training operations in the Baltic States that have taken me on numerous occasions to Latvia, Estonia, and Lithuania, as well as Moscow. This provided me with firsthand observations of the way the KGB and MVD operated. Once on a tour in Tallinn, Estonia, my men and I were warned that some individuals had approached the staff of the government hotel with questions about us. That evening we were taken for dinner at the Viru Hotel. Since we had an early start the next morning, I left two of my men, Mickey and Anders, in the restaurant and returned to the hotel to pack.

When my two cohorts left a couple of hours later, they were ambushed by 10 Russian mafia. They thought we weren't armed because no holsters were visible when we had taken our jackets off in the restaurant. When the thugs made their move, the Russian police and hotel security turned around and walked out, leaving Mickey and Anders on their own. They drew their weapons as the Mafia came for them, overcoming the language barrier thanks to Mr. Smith & Wesson and Mr. Glock, who were well hidden. In a short spate of time the mafia looked like pilgrims, first on their knees and then on their bellies. This was the first of three meetings our teams would have with the mafia. On this occasion no shots were fired, but on subsequent occasions the score would be IBA 4, Mafia 0.

PREFACE

Various Baltic States antiterrorist teams being trained by Jim Shortt before and after independence. Photos courtesy of Jim Shortt

KGB ALPHA TEAM TRAINING MANUAL

I have mused since that the attacks were simply bad luck for the mafia picking on armed tourists. Senior Baltic interior ministries' personnel have told me that that wasn't the case. IBA was the target, and the operations were sanctioned by the KGB in Minsk, where the KGB moved all its Baltic operations after the failed coup and where they still target the Baltic States,

In January 1991, an attempted coup took place in the Baltic States. Two airborne divisions specifically targeted on the Baltic States, 7th Guards Airborne with its headquarters at Kaunas in Lithuania (supported by the Spetsnaz brigade at Mariinogorko) and 76th Guards Airborne at Pskov (supported by the Spetsnaz NCO training brigade at Pechory). The Spetsnaz units were tasked with seizing important buildings such as government offices, the president's palace, and TV and radio stations. The commander of the 7th Guards Airborne, Col. Vladimir Federenko, had some sympathy for the nationalists. He realized that his officers and men were being used in a political game. (Furthermore, conditions for the military men had generally deteriorated. Their wives and children had were living in communal buildings with blankets hung on ropes for walls.) As tension with the nationalists governments increased, food for the troops—traditionally supplied by the civil administration of the republics—became short. Moscow deliberately created tensions in the lives of the Soviet military personnel and *zampolits* directed the anger at the nationalists. When Baltic youths refused to accept conscription into the Soviet forces, the paras were sent in with military police units to hunt them down.

The situation escalated when a series of mysterious bombs started to explode in Latvia, targeting a statue of Lenin, Communist party headquarters, KGB headquarters, a military hospital, and the military institute. The Latvian police arrested a young Russian, but the explosions continued—the work of Interfront assisted by the republic's KGB Alpha team. After the success of the Alpha project in Moscow, it was decided to set up Alpha teams in KGB headquarters in every republic, since the country was too vast for just one team to be able to respond, especially if there was more than one mission assigned. The teams, comprised of 12 to 15 men, trained in Moscow and by 1990 existed in Tallinn, Riga, and Vilnius.

On 12 January—while the world was busy watching the Middle East, as it had been in 1956 when the Soviets attacked Hungary—the 7th and 76th Guards Airborne moved on the Baltic States. At the top of the list was Lithuania. The 7th Guards moved in and occupied the OMON headquarters in Vilnius and then attacked the TV station. BMD-1 and BMD-2 tanks crushed unarmed demonstrators beneath their tracks as soldiers fired into the demonstrators. But when the paratroopers, sickened by the slaughter, held back, KGB Alpha personnel based in Vilnius and dressed in VDV uniforms led the final assault. The attack and its brutality were recorded on video and shown on television screens throughout the world.

Members of the MVD Spetsrota train in arrest and search techniques. Photo courtesy of Novosti Information Agency

Spetsnaz from Mariinogorko made a half-hearted attempt to seize the Journalists' Union building and the telephone exchange, firing live rounds above the heads of the crowd and at one point fixing bayonets to get through them. Eventually the 50-man unit retreated, leaving the paras to confront the crowd. At Lithuanian Defense Forces headquarters, KGB Alpha team members stormed the building, using stun and smoke grenades. The bodyguard team for President Landsbergis barricaded the parliament building; issued weapons, ammunition, and gas respirators; and awaited an assault. The inner protection core was made up of Lithuanian IBA members who a month before had come to Tallinn for training. Now in body armor, they watched, ate, and slept with their AKS-74s always at hand as they shadowed Landsbergis' every move.

On 15 January, under pressure from the rest of the world, Gorbachev tried to rein in the Alpha teams, much to the dismay of the conspirators in Moscow. The next day, OMON forces based at Vecmilgravis Bridge in Riga opened fire on a passing car, killing the driver for no apparent reason. This senseless murder heightened tensions. The Riga OMON, assisted by Soviet troops, seized the Latvian police academy. They then moved to the Latvian Interior Ministry with BTR-70 ARCs and tried to ram blockading vehicles unsuccessfully. The Latvian Interfront declared itself a "National Salvation Committee" and the new government of the republic. OMON fired some irritant gas and left.

This action came after Col. Gen. Fyodor Kuzmin, the Baltic District Defense commander, ordered all Latvian police (MVD militia) to hand in their weapons and ammunition. Under the Soviet system, Soviet police have their weapons and ammunition lent from the army—they do not ultimately belong to the republics' interior ministries. President Gorbachev insisted he had not authorized the action in Vilnius, which left 14 dead and 230 wounded.

On 10 January 1991, the OMON unit at Riga was joined by a sniper team from the KGB Alpha team. The operation was a diversionary attack on the Interior Ministry headquarters in Riga, while Alpha team snipers were dropped off on Padomju

Boulevard at the back of a park facing Raina Boulevard, the site of the Interior Ministry. They ascended a small hillock and positioned themselves within a stone ornamental fort so that they had a clear shot down the short R. Endrupa Street at the side of the ministry. Located at 1 R. Endropa Street is the Ridzene Latvian Government Hotel, where Prime Minister Ivars Godmanis was playing host to a Polish trade delegation at a formal dinner in the restaurant on the mezzanine. When the firing started, the bodyguards rushed the prime minister out of the building, but he was shot by a stray round from an Alpha team 7.62mm sniper rifle.

Afterward, OMON claimed it had gone to the Interior Ministry to complain about the alleged rape of the wife of an OMON man. Video footage shows that they stormed the building, firing indiscriminately, killing a Ukrainian police officer, and forcing people to dive for cover. The Alpha team fired some shots for effect. Below them in the park, a TV film crew saw the flash and movement and went to investigate. Two of the crew were shot, including the cameraman, who kept the camera rolling as he died. A police lieutenant rushed through the park to the cries for help, and as he approached the small footbridge, he too was shot by the sniper team's security party. The prime minister's bodyguards, suspecting an ambush, took the VIPs out a back entrance.

The prime minister reached his office safely and telephoned Marshal Yazov for an explanation. Yazoz said he knew nothing of the affair. The OMON group led by Lt. Alexander Kuzmin, an Estonian-born Russian, numbered approximately 100. The Interior Ministry contained the deputy minister and 14 others.

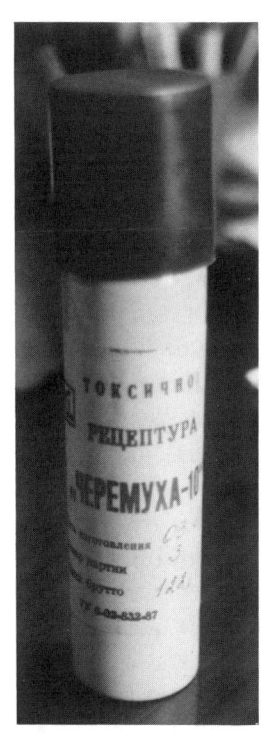

Soviet-issue "cherry" or irritant gas given to KGB and MVD units. Photo courtesy of Jim Shortt

Next, the OMON unit attacked the Ridzene Hotel, firing wildly at the glass front. Then shots were fired at the mezzanine. The bullet holes can still be seen in the glass balcony. The casualties from the OMON/Alpha operation numbered two police and two civilians dead and two police and seven civilians wounded.

The attacks on the customs posts in the three Baltic republics became the center piece of OMON action until the Moscow coup. On 20 March, the Vilnius OMON opened fire on a bus of unarmed Lithuanian border guards returning from their shift and wounded three men. The OMON beat up two of the wounded and took them to a local Soviet military garrison. Two days before, the Vilnius OMON had seized Lithuanian Defense Minister Audrius Butkjavicius in the street. In mid-May, more customs posts in Lithuania were attacked, and on the night of May 22, the Riga OMON attacked and destroyed customs offices in Latvia. On the same night, armed men in civilian dress (from Interfront's paramilitary arm) attacked Estonian customs offices.

On 30 May, the Riga OMON was placed under investigation for criminal acts by the USSR procurator general. Throughout the summer attacks on the Baltic States customs post continued. On 7 June armed men in civilian clothes attacked and destroyed an Estonian customs post at Luhamaa. On 9 June another Estonian customs port near Narva was attacked and destroyed. Two customs official were kidnapped and taken into the Russian Republic, where armed Russian MVD militia arrested the terrorists, who turned out to be members of the Interfront (Estonia) paramilitary unit.

Following this attack the Estonian government decided to arm its border guards. But before this policy could be put in to operation, unidentified armed and uniformed man attacked and destroyed border posts in Estonia, Latvia, and Lithuania. This was followed by an attack by Riga OMON on the customs post in the capital. On 18 June unidentified armed men again attacked posts in Latvia and Lithuania.

On 26 June, the Latvian OMON commander, Major Makutinovich, turned up in Lithuania to lead an OMON raid on the central telephone exchange. This effectively cut Lithuania off from the world but also threw international air-traffic control lines into jeopardy. Coordinated with this attack, armed men attacked two electrical energy transfer stations. Their action placed the Ignalina nuclear power station at risk.

In Riga, a former OMON soldier, who had resigned in disgust after the attempted assassination of the prime minister, was found murdered. The soldier had contacted former colleagues in the Latvian Interior Ministry and told them what was happening within the OMON, providing photographs and membership records for the unit. The scene-of-crime officers believe the slain soldier knew his attacker; apparently he had been called over to a car, a pistol was shoved in his eye, and the trigger pulled—a technique shown in this manual. Murder had become the name of the game.

In Moscow, 300,00 demonstrators protested against the action in the Baltic States and called for Gorbachev and Yazov to resign. The Moscow OMON and the Felix Dzerzhinsky Division waited with other Interior Ministry troops in the side streets behind the GUM department store in case of trouble. On the USSR's western border, the 103rd Guards Airborne Division, under control of the KGB Border Guards Directorate, was deployed. Among other things, the division seized film and videotape from journalists who had come from the Baltic States. On 27 January 1991 the Lithuanian OMON was at it again, this time destroying two Lithuanian customs posts.

The prosecutor general of the USSR, General Nikolai Trubin, later ruled that OMON's actions in Lithuania and Latvia were illegal and the members should face prosecution, but the USSR MVD and KGB did not assist his enquiries, claiming that OMON was not their responsibility. Yet Czeslaw Mlynnik, who participated in the Spets group's attack on President Amin's Palace in December 1979, led the Latvian OMON. In October 1990, just before the attack, he was a senior lieutenant. In January he made captain, and later he became a major—all within six months. His coconspirator was Alfred Rubiks, the former Latvian CPSU boss. In the months before the August coup, Major Makutinovich would be a regular and welcome visitor at 46 Ogaryova Street, Moscow—the MVD headquarters for the USSR.

In February 1991, the Estonian MVD opened its first police school at Paikus, south of the port of Parnu. I was asked to return that spring to Estonia to retrain the bodyguards, advise on future equipment purchase, teach a CQB course at the new school, and train members of the police reserve.[33] I returned to Estonia on my own, and the KGB customs service played games with the paperwork. This time the papers for my weapons were in order, but I required the signature of tho KGB boss for Estonia, Rein Sillar, and he could not be contacted. After conferring with the secret service officers who met me, I decided to let the KGB hold the weapons until Sillar could be located, and in the meantime the Estonian police gave me a Makarov and ammunition to carry for my own safekeeping.

The training took place, and I made plans to return with a four-man IBA team in June. At Pirkus I arrived in the president's Chaika limousine, lent for the journey. Weapons handling and firing of AKS-74U and Makarov (PM) pistol training took place in CQB-mode, followed by *rukopashnyi bol,* arresting techniques, and prisoner handling. At this point, we openly referred to the terrorists as Lithuanian and Latvian OMON, Estonian Interfront activists, and Russian mafia. Boris Pugo, the USSR MVD minister, had gone an record to say that Estonia had the worst crime rate in the USSR. Of course it did—he and his KGB colleagues had exported all their criminal gangs to the country. In January, at the height of the Baltic crisis, two Swedish trade union leaders, Sartil Whinberg and Ove Froderiksson, had been kidnapped, robbed, and murdered. This had the effect of dissuading other foreign visitors from coming to invest in the Baltic States, for a while.

At the end of May in 1991, the USSR Central Sports Committee based at Dynamo in Moscow invited me to fly to Moscow and discuss training troops. I was asked not to bring any weapons, as these would be provided for me. I was met before I reached customs by Josif, a member of the committee and an interpreter from Moscow University. Over lunch, I met "former" KGB officers, including Anatoli, a veteran of the Afghan War and Osnaz. Arrangements were made for me to travel to the Crimea,

PREFACE

Estonian special unit at Tallinn Airport being trained by Jim Shortt to handle hijack scenarios. The troops carry equipment taken from KGB Alpha Teams. Photo courtesy of Jim Shortt

even though my visa permitted travel in Moscow only.

Josif and I took a reserved sleeping compartment and some caviar and set off by rail for Simferopol in the Crimea. We passed through Tula and the Ukraine, and the next morning we entered the Crimea. Our carriage attendant, a young woman, had alerted railway police that a foreigner was on board without the correct visa. At Simferopol station, Nikolai, a major in the KGB, met the train with a giant of a man, another KGB officer named Sasha, and a KGB interpreter. The major took out his red KGB ID card, and the car attendant looked like she would shrivel to dust on the spot. He spoke firmly to her, and she just nodded repeatedly.

We loaded in to a KGB microbus, a VW look-alike called a Latvia. This took us to the Crimean KGB headquarters. Opposite was the Dynamo Sports Center and Dynamo Hotel reserved for KGB use. A suite was provided with bedroom, bathroom, sitting room, and hall. I was impressed. Over the next days KGB agents Nikolai, Sasha, and Volodya were my tourist guides. Before I left I ran a short course at the DOSAAF center for KGB person-

KGB Alpha Team target shows little girl being held by hostage. Photo courtesy of Jim Shortt

PREFACE

These photos were taken during training sessions conducted by Jim Shortt and other IBA personnel at the Soviet Airborne Forces Academy from February through August 1989.

KGB ALPHA TEAM TRAINING MANUAL

PREFACE

nel on instinct shooting and hitting partly obscured targets where hostages are involved. The KGB Alpha team targets were produced, showing a pistol-wielding desperado holding a little girl in front of him. The KGB people were suitably happy, but then came the crunch: they asked me to stop working with the Baltic States' governments. After all, the KGB said, the Baltic States were just boys, amateurs. The KGB hinted that if I stopped, then courses for

OMON and other specialist MVD and KGB units could be arranged. I politely declined.

· · · · ·

This manual was used to train special assignment personnel who now work for a Russian nation that still wishes to dominate or influence the area of the former Soviet Union, or who work for Russian criminal gangs. I do not believe the West's problems with the former Soviet Union finished with the coup's failure. This manual is a product of covert active measures. When I trained anti-Spetsnaz special units within NATO and for neutral governments, such as Sweden, we weren't allowed to identify the threat as being Soviet. NATO spoke of "Orange Forces," and the Swedes referred to "*Sabotageforband.*"[34] The pictures of Osnaz in this manual killing U.S. Special Forces and U.S. 1st Cavalry Division personnel show that the Soviet Union suffered no such restrictions in their training.

NOTES

[1] Spetsnaz—acronym of *voiska spetsial'nogo naznacheniya* forces of special assignment or special-purpose forces
[2] Naval Infantry—similar to U.S. Marine Corps
[3] MVD—*Ministerstvo VvnutrennikhDel*, Ministry of Internal Affairs
[4] KGB—*Komitet Gosuderstyennoi Bezopasnosti*, Committee for State Security
[5] CHON—*chasti osobogo naznacheniya*, special assignment detachments
[6] The KGB was still hunting the Estonian Forest Brothers well into the 1960s.
[7] OMSBON—*Otdelnyei motostrelkovye brigady osobnnogo naznacheniya*, or independent special purpose motorized brigades
[8] Osnaz—an abbreviation of *osobogo naznacheniya*, another specialist designation
[9] Communist Party of the Soviet Union
[10] *Smert'Shpionam*—Death to Spies
[11] GRU—*Glavnoe Razvedyvatel'noe Upravlenie*, the central reconnaissance directorate of the Defense General Staffs
[12] GTO—*Gotov k Trudei Oborone*, Ready for Labor and Defense of the USSR program of the All Union Sports—Technical
[13] DOSAAF—*Dobrovol'noe Obshchestro Sodeistriya Armii*, Aviatsii i Flotu, Voluntary Society for Cooperation with the Army, Air Force, and Navy
[14] "Summer Lightning"

[15] All-Union Leninist Communist Union for the Young
[16] "Little Eagle"
[17] *Voennosportivnyikompleks*—military sports activities
[18] In Russian, *kholodnye oruzhiye*
[19] *Samooborona bez oruzhiya* in full
[20] Equivalent to officer cadets
[21] Spetsrota
[22] *Krapovye berety*
[23] *Gruppa zakhvata*
[24] *Imitatsiya*
[25] *Zarga*
[26] No relation to the Italian variety
[27] *Otryady militsii osobogo naznacheniya*—police battalion of Special assignments
[28] By K.G. Andreev and M.Y. Davidov
[29] Russian term for Afghan War veteran (plural, *afghantsy*)
[30] A comfortable, two-bedroom suite (I stayed in one in Ryazan), but not up to the standard of a U.S. enlisted person's quarters
[31] Boris Pugo, former Latvian KGB chief
[32] The Azerbaijani OMON numbered 5,000 men.
[33] A special (not reserve or part-time) unit held in reserve for antiterrorist work that has regular officers, mostly ex-Spetsnaz
[34] Sabotage groups

TRANSLATOR'S NOTE

It is not often that one outside of intelligence or military circles has the opportunity to read an official armed forces publication from the ex-Soviet Union. I am glad to have had the chance to work on this one.

It is generally a safe assumption that the language and style of military manuals the world over are stilted, limited, and repetitious. I was not disappointed in this by *KGB Alpha Team Training Manual.* I have rendered the original into English as faithfully as possible, including the retention of a few russianisms or sovietisms.

The much beloved phrases *moral no-psikhologicheskii* and *moral no-politicheskii* have no exact equivalents in English, but are (or were) extremely important in party and military doctrine. They are given in this translation as "moral-psychological" and "moral-political." They have little to do with morality, but everything to do with morale. The term "cold weapon(s)" has been kept: it is very descriptive (and authentic) and avoids endless repetition of "edged/pointed weapons" or some such.

I have not tampered with archaisms (e.g., knocking an enemy off a horse), or some of the alarmingly simplistic, not to say optimistic, how-tos for certain actions. A few vague descriptions in the original had to be finessed by the use of "hand/arm" or "feet or legs," for instance. Hand and arm have the same word in Russian, as do foot and leg. Finally, the opportunity to translate this manual has ensured that I will never forget the Russian word for crotch (for those interested, it is *promézhnost*).

Peter Bercé
Translator
Washington, D.C.

This manual offers a study of special physical education and some questions on concurrent monitoring of military personnel's state of health and application of means to restore physical strength after exertion.

The manual is meant for physical education specialists and for commanders and instructors of units with increased combat readiness.

The author wishes to thank comrades S.A. Golov, A.A. Nabokov, N.A. Zubkov, A.N. Kharin, and the unarmed combat specialists of military units 33965 and 35690.

Editorial and production supervision was by V.P. Simonov.

Art production was by N.A. Bil'din.

Foreword

This manual contains practical material on the movement of troops in varied settings, overcoming obstacles and unruined positions, personal/hand-to-hand combat that employs feints and ruses and diverse physical objects, and gives information for monitoring health and ways of restoring it after heavy physical exertion.

The manual's purpose and contents are a response to the need for training armed forces personnel under conditions that are as close as possible to those of true combat operations. Superior combat effectiveness of the personnel in varied situations is the determining factor in carrying out missions. Such effectiveness ensures a high level of political and moral-psychological preparedness and physical toughness. Under conditions of actual danger that produce significant psychological stress, the outcome of combat operations will depend fully on skillful and decisive action. Achieving the highest results promotes the skillful use of personal weapons and the sure application of the different forms of personal combat in swiftly changing situations.

Troops' moral-political preparation and psychological conditioning become real in the course of combat, political training sessions, and special physical training. High moral-psychological qualities are shaped in all ways by military service and its system of occupation assignments.

In a study of exercises, activities, and techniques that entail risks, it is important, in the interests of the trainees' psychological well-being, to think through beforehand a system for protection against injuries.

The manual's instructional material examines the effects of physical exertion on personnel. Endurance is the fundamental quality, the basis of the soldiers' special physical training. The manual presents an array of measuring methods that permit a fairly accurate assessment of the levels of development of this quality of endurance under combat conditions. The manual's contents are enhanced by a considerable amount of graphic material. Some aspects of special physical training are shown in Figure 1.

Physical education is an ingredient of Communist education.

FIGURE 1. ASPECTS OF SPECIAL PHYSICAL TRAINING

CHAPTER 1
THE FOUNDATIONS OF SPECIAL PHYSICAL TRAINING

This [physical education] strengthens the health of the Soviet people and aids their all-around development. Physical education is one of the important means of preparing the nation for work activity and for the defense of the Motherland. The Party's Program presents the task of educating the new person ["new Soviet Man"], harmoniously combining in him spiritual wealth, moral purity, and physical perfection.

The ideological foundation of our physical education system makes manifest the Marxist-Leninist doctrine of the unity and indissolubility of the system of mental and physical growth. Data from the social and natural sciences constitute the scientific-methodological basis of that system, which is also the only national program for physical education in existence.

Physical training, as one kind of physical education, is a process of perfecting the motor capabilities of an individual, keeping in mind the specific traits of that individual.

The principal recipients of physical training, which is made clear in the manual, are armed forces personnel. The training is governed by programs confirmed by orders of the relevant ministries and administrations.

Physical training stands out as a means of losing excess weight, while raising combat readiness. Aside from these, the training methods can bring out some special, outstandingly pro-

BASIC ELEMENTS OF SPECIAL PHYSICAL TRAINING

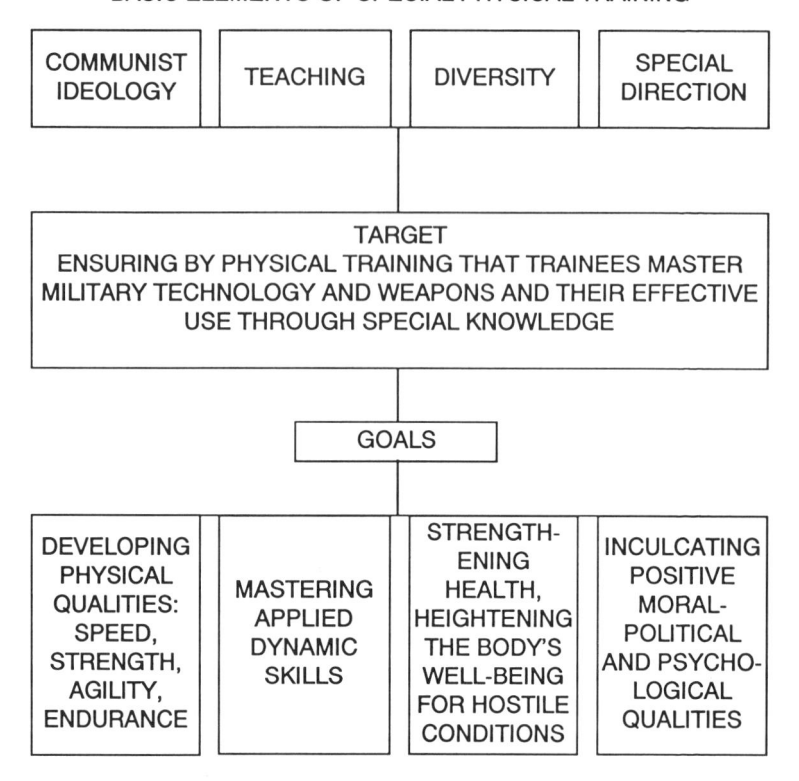

COMMUNIST IDEOLOGY	TEACHING	DIVERSITY	SPECIAL DIRECTION

TARGET
ENSURING BY PHYSICAL TRAINING THAT TRAINEES MASTER MILITARY TECHNOLOGY AND WEAPONS AND THEIR EFFECTIVE USE THROUGH SPECIAL KNOWLEDGE

GOALS

DEVELOPING PHYSICAL QUALITIES: SPEED, STRENGTH, AGILITY, ENDURANCE	MASTERING APPLIED DYNAMIC SKILLS	STRENGTHENING HEALTH, HEIGHTENING THE BODY'S WELL-BEING FOR HOSTILE CONDITIONS	INCULCATING POSITIVE MORAL-POLITICAL AND PSYCHOLOGICAL QUALITIES

FIGURE 2. BASIC ELEMENTS OF A PHYSICAL TRAINING SYSTEM; ITS TARGETS AND GOALS

fessional qualities, which can then be more quickly recognized and developed for the needs of specific branches of the service.

An objective of physical training is the methodicalness of building training for personnel and directing such a process.

Within the training of armed forces members are the selection and application of the means, methods, and forms that, at an operational level, guarantee the general and specific goals of each service branch or occupation. Physical exercise, sports, and restoration of natural strength serve as the means. The direction of physical exercise, the utilization of various sports, and maintenance of natural strength are the methods. Systematic use of the methods and specified ways of influencing individuals' constitutions in toto, along with pedagogical techniques, are the body of the methods. Special physical training is closely tied to the troops' intellectual and moral-aesthetic instruction and technical education in the process of mastering different training and disciplines.

Achieving physical goals after intensive mental activity is an active way of relaxing and safeguarding the lasting acquisition of knowledge. In planning exercises, the principle of alternating theoretical with practical drills requires a workable schedule. The teaching must be systematic.

Special physical training, while heightening combat readiness and discarding anything superficial, hardens the trainees through a plan of morale/willpower, producing psychological stability under unfavorable circumstances and professional responses.

Such results are achieved through specially chosen physical exercises, activities, devices, and myriad approaches.

Figure 2 presents schematically the basic elements of physical training and its goals and aims.

Through physical exertion, internal changes and shifts of an anatomical-physiological character appear, which manifest themselves in a human organism under the influence of this exertion. The shape physical training takes is the stuff of drills, morning calisthenics, different types of independent work, the training done in sports training units, and so on (Figure 3).

The basic forms of special physical training determine the activities for all personnel.

Special physical training stays on its target with the help of effective means, methods, and forms to ensure the superior development of special physical and psychological qualities and applied skills.

Special physical training has:
- General goals:
  — Developing strength, agility, and endurance;
  — Mastering a range of applied movement skills such as getting through obstacles and water barriers, and skiing;
  — Strengthening health, improving indexes of physical development, hardening and raising the organism's defense against hostile factors;

FORMS OF SPECIAL PHYSICAL TRAINING

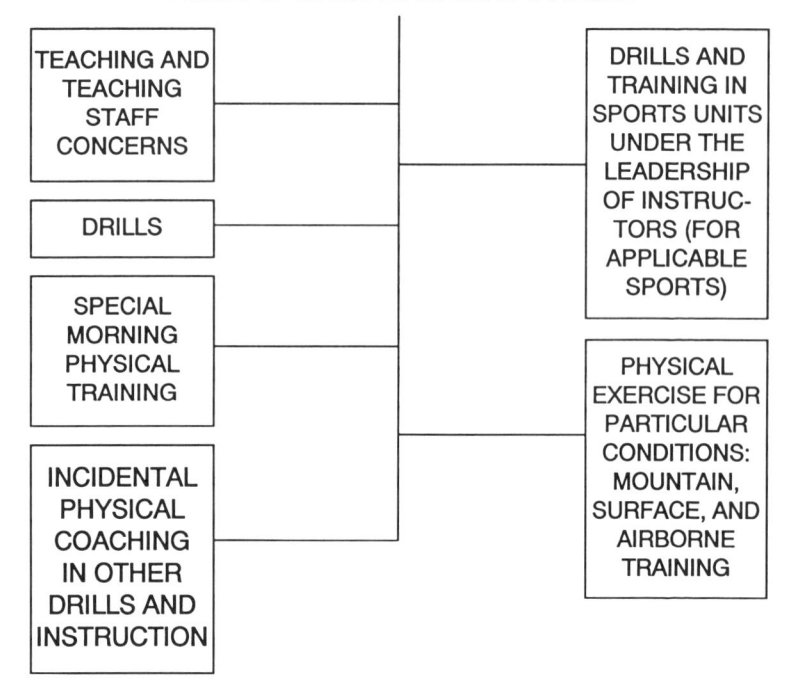

```
┌─────────────────┐                    ┌─────────────────┐
│ TEACHING AND    │                    │ DRILLS AND      │
│ TEACHING        │────────┐           │ TRAINING IN     │
│ STAFF           │        │           │ SPORTS UNITS    │
│ CONCERNS        │        │           │ UNDER THE       │
└─────────────────┘        │           │ LEADERSHIP      │
                           │           │ OF INSTRUC-     │
┌─────────────────┐        │           │ TORS (FOR       │
│ DRILLS          │────────┤           │ APPLICABLE      │
└─────────────────┘        │           │ SPORTS)         │
                           │           └─────────────────┘
┌─────────────────┐        │
│ SPECIAL         │        │           ┌─────────────────┐
│ MORNING         │────────┤           │ PHYSICAL        │
│ PHYSICAL        │        │           │ EXERCISE FOR    │
│ TRAINING        │        │           │ PARTICULAR      │
└─────────────────┘        │           │ CONDITIONS:     │
                           │           │ MOUNTAIN,       │
┌─────────────────┐        │           │ SURFACE, AND    │
│ INCIDENTAL      │        │           │ AIRBORNE        │
│ PHYSICAL        │────────┘           │ TRAINING        │
│ COACHING        │                    └─────────────────┘
│ IN OTHER        │
│ DRILLS AND      │
│ INSTRUCTION     │
└─────────────────┘
```

FIGURE 3. FORMS OF SPECIAL PHYSICAL TRAINING

— Forming moral-political and psychological qualities.
• Specific goals:
— Mastering techniques of taking prisoners, silently eliminating sentries, securing and transporting prisoners;
— Developing skill in acting as a unit in attack groups and capture groups;
— Building knowledge and skills to defeat an enemy in single combat, without using firearms, utilizing sambo, boxing, karate, and judo.

Ways of achieving these goals are presented in Figure 4.

THE FOUNDATIONS OF SPECIAL PHYSICAL TRAINING

WAYS OF ACHIEVING THE GOALS OF SPECIAL PHYSICAL TRAINING

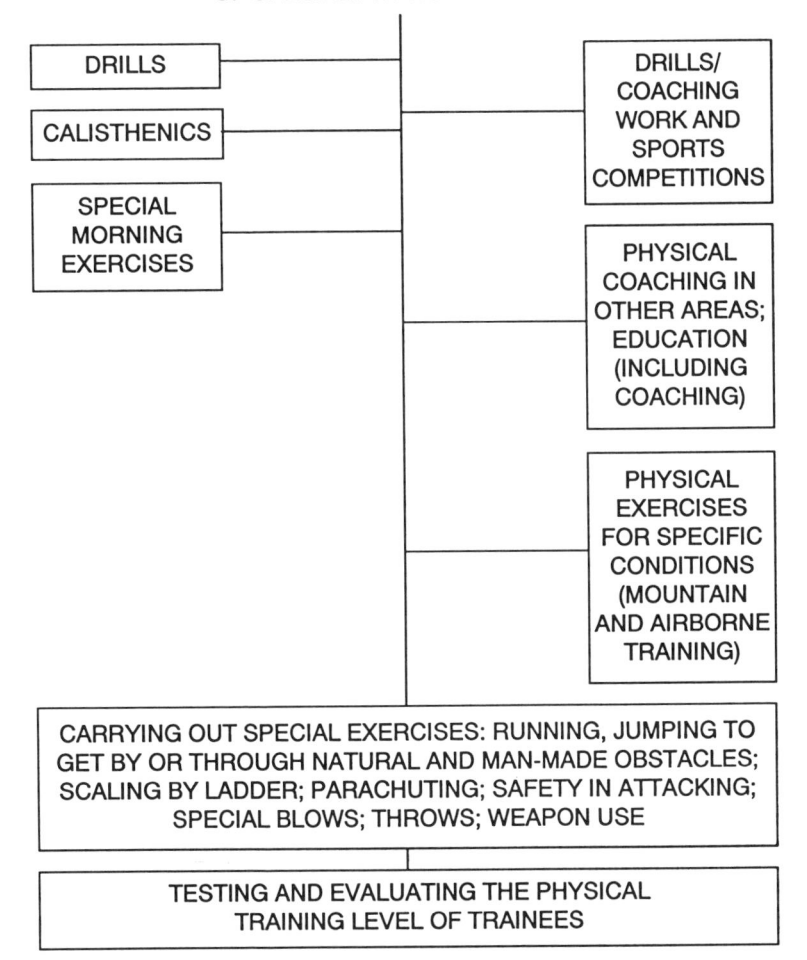

- DRILLS
- CALISTHENICS
- SPECIAL MORNING EXERCISES
- DRILLS/COACHING WORK AND SPORTS COMPETITIONS
- PHYSICAL COACHING IN OTHER AREAS; EDUCATION (INCLUDING COACHING)
- PHYSICAL EXERCISES FOR SPECIFIC CONDITIONS (MOUNTAIN AND AIRBORNE TRAINING)

CARRYING OUT SPECIAL EXERCISES: RUNNING, JUMPING TO GET BY OR THROUGH NATURAL AND MAN-MADE OBSTACLES; SCALING BY LADDER; PARACHUTING; SAFETY IN ATTACKING; SPECIAL BLOWS; THROWS; WEAPON USE

TESTING AND EVALUATING THE PHYSICAL TRAINING LEVEL OF TRAINEES

FIGURE 4. WAYS OF ACHIEVING THE GOALS OF SPECIAL PHYSICAL TRAINING

THE FOUNDATIONS OF SPECIAL PHYSICAL TRAINING ORGANIZATION

The teaching process in special physical training (SPT) encompasses planning material-technical assurance, physical conditioning to facilitate training, and periodically monitoring the speed of mastering the training course's content.

Planning is built on a foundation of directives from the highest administrative bodies.

Organization of the body of instruction takes into account trainees' specific activities:
- The tie between the training process and the nature of the trainees' future assignments
- The systematization and regulation of the body of instruction
- The evenness of the distribution of physical demands in a week's course and
- The condition and level of physical development of the trainees, as well as their material-technical development

PRINCIPLES OF INSTRUCTION

Instruction implies the regular transfer of knowledge to trainees, producing in them abilities and skills with the goal of developing and realizing their physical capabilities. Instruction, together with testing, is a unitary pedagogical process, which is founded on the following principles:
- Party discipline and knowledge
- Consciousness
- Action
- Demonstrability
- Systematicness
- Gradualness and accessibility
- Lasting acquisition of knowledge

Party Discipline and Knowledge
This is realized in the process of continual clarification of those goals that trainees must attain after they have mastered the whole range of SPT methods and can utilize their latest technical accomplishments.

Consciousness
This is made real on the basis of trainees understanding the

necessity of mastering all means of SPT, a clear presentation of the exercises or activities and their effects on the human body. In consequence, there has to be development of an ability to analyze success and failure in performing exercises, activities, and techniques.

Action

This demands a clear explanation of the system of grading and encouragement to motivate the trainees, as well as in any independent training.

Demonstrability

This is realized in two ways:
- Giving demonstrations with oral descriptions
- Using obvious examples

Systematicness

This dictates a defined system of instruction. New material must be a continuation of the old, and its use systematic. Material, therefore, should be arranged methodically and used that way for SPT.

Gradualness and Accessibility

This is realized in the results of [the trainees'] passage through training, from the simple to the complex, facing a gradual increase in physical training demands.

Lasting Acquisition of Knowledge

This means much repetition of the physical exercises in various combinations and circumstances, with obligatory verification and rating of the knowledge acquired.

All the principles are valid in teaching because they are interlinked, and they manifest themselves in a single kind of training.

There are three groups of [teaching] methods:
- Oral—explaining and describing the body of drills and activities, meeting with participating trainees
- Demonstration—using visual or aural presentation, to

include accurate illustrations, diagrams, and transparencies
• Practical—encompassing various repetitions, i.e., exercises, all of one kind or different, in simplified conditions or difficult ones

Aside from these instruction methods, there are such methodological techniques as testing, mutual help, safety, and so on. Physical training instruction comprises a methodical sequence of stages.

First Stage: Familiarization

The goal is to create the right presentation of exercises, activities, and techniques. Here, this means clearly and briefly naming an exercise or technique, personally demonstrating it in full, describing its effects on the human body when it is performed (or adapted for practice), demonstrating it again (but a bit at a time), and, in passing, explaining the technical aspects slowly.

Second Stage: Learning

The goal is to shape the ability to perform techniques and actions. The whole [process] is accomplished, bit by bit, with the help of the trainees and their commanders. The instructor provides the trainees with the opportunity to slowly complete an exercise in its entirety, to work on it independently at a gradually increased pace, and, at the end, forces the trainees to perform in quick (military) time. In all this, good performance and mistakes are noted and ways to improve are laid out.

Third Stage: Training

The goal is to sharpen skills in performing exercises in varied circumstances. At this stage, the trainees perform exercises from different starting positions, while moving, or in combination with a mix of other exercises. Furthermore, the trainees carry out an exercise correctly and quickly in a competitive environment and combine the exercise with those learned earlier. The complete step-by-step process is presented in Figure 5.

In the training-teaching process, the combined exercise method is preferred. Two or three parts of physical training (e.g., techniques of movement, getting by obstacles, or elements of

THE FOUNDATIONS OF SPECIAL PHYSICAL TRAINING

PROGRESSIVE TRAINING BY EXERCISES, ACTION, OR INDIVIDUAL ACTIVITIES

FAMILIARIZATION GOAL: CREATE PROPER PRESENTATION	LEARNING GOAL: SHAPE PERFORMANCE SKILLS	TRAINING GOAL: SHARPEN SKILLS FOR PERFORMANCE UNDER VARIED CONDITIONS
BRIEFLY AND CLEARLY NAMING ACTIVITIES, TECHNIQUES, EXERCISES. DEMONSTRATE THEM IN QUICK TIME.	MEANS: (1) AS A COMPLETE FUNCTION; (2) BY THE NUMBERS, SEGMENTS; (3) WITH EXERCISE LEADERS' ASSISTANCE	PERFORMING TECHNIQUES, EXERCISES, ACTIONS IN VARIED STARTING POSITIONS; IN COMPLEX CIRCUMSTANCES, WHILE MOVING
THE TACTICAL USE EXPLAINING	PERFORMING THE WHOLE FUNCTION AT AN INCREASING TEMPO	PERFORMING EXERCISES IN COMBINATION WITH OTHER KNOWN ACTIONS OR TECHNIQUES
DEMONSTRATING THEM AGAIN, BUT SLOWLY, BY PARTS; EXPLAINING TECHNICAL ASPECTS	ASSIGNING SHORT, INDEPENDENT PERFORMANCES OF ELEMENTS OF THE WHOLE, CORRECTING MISTAKES	PERFORMING EXERCISES, ACTIONS, TECHNIQUES WITH SPEED AND ACCURACY IN A COMPETITIVE ENVIRONMENT
SEPARATING AND TURNING ATTENTION TO MAIN ELEMENTS OF EXERCISES, ACTIONS, AND TECHNIQUES	PERFORMING EXERCISES BEING TAUGHT, WITH ACTIVITY AT A MILITARY (QUICK) PACE	PERFORMING A COMPLEX OF, OR IN COMBINATION WITH, EARLIER LEARNED TECHNIQUES, ACTIONS, AT A MILITARY PACE

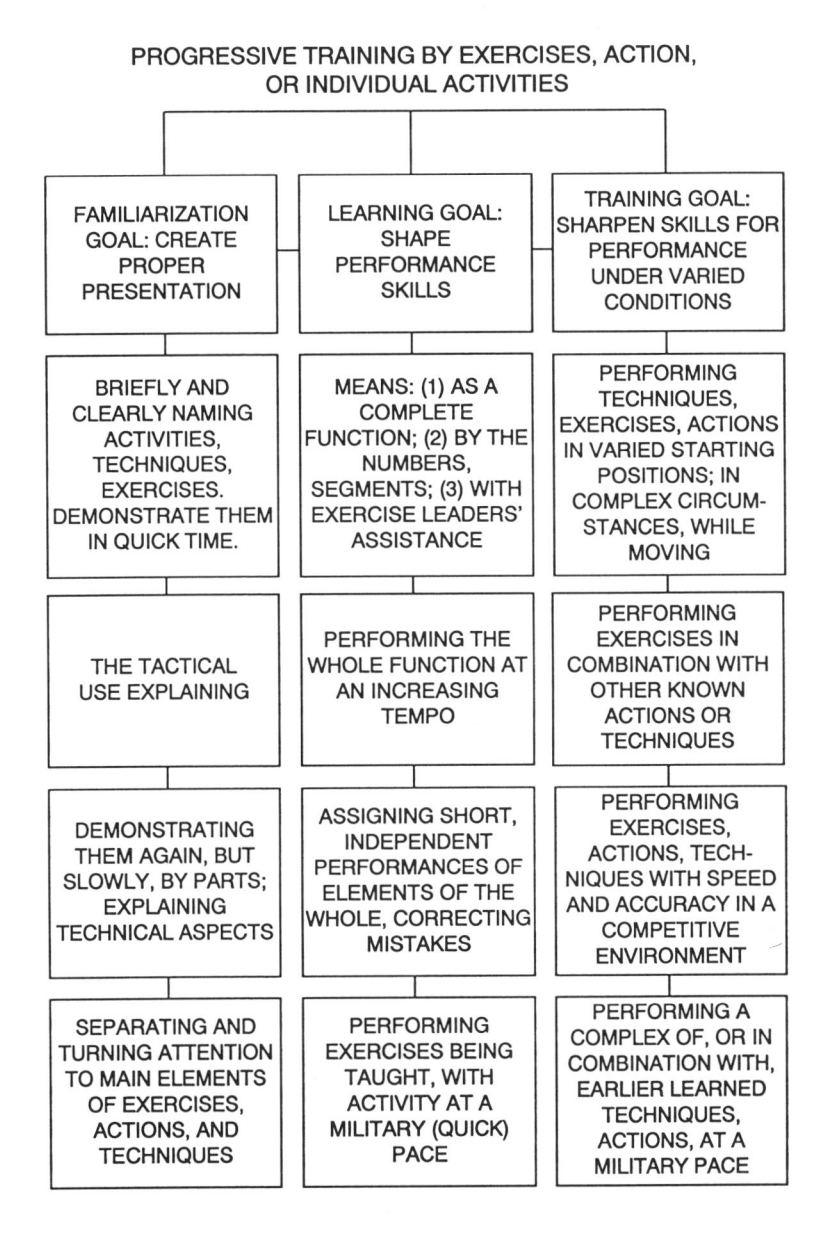

FIGURE 5. PROGRESSIVE PROCESS OF TEACHING PHYSICAL EXERCISES

KGB ALPHA TEAM TRAINING MANUAL

NO.	TEST NAMES	KIND OF EXERCISE OR METHOD OF PERFORMANCE	RATING EXCELLENT	RATING GOOD	RATING SATISFACTORY	REMARKS
1	SKIING	5-KM. RUN	26 MIN.	27 MIN.	29 MIN.	WITH EQUIPMENT, RIFLE, FLAK JACKET
2	LIGHT ATHLETICS	3-KM. RUN	13 MIN.	14 MIN.	15 MIN.	(6-7) KM.)
3	POWER GYMNASTICS. PULL-UPS	OVERHAND GRIP	8-7 TIMES	7-6 TIMES	5-4 TIMES	(6-7 KM.)
4	ROPE CLIMBING	WITHOUT USING THE LEGS	5 M.	4 M.	3 M.	(6-7 KM.)
5	OBSTACLE COURSE	SERIES OF 14 OBSTACLES, TRAPS	4 MIN.	4 MIN. 30 SEC.	5 MIN.	FOR THE FIRST OBSTACLE A 30-SECOND HANDICAP
6	THROWING EDGED/ POINTED WEAPONS AT A TARGET	FIELD DISTANCE: 6 PACES, CHEST AS TARGET	5 HITS	4 HITS	3 HITS	
7	FORCED MARCH IN A FLAK JACKET OR CARRYING 10 KILOS					
8	COMBINES EXERCISES TEST 2		5 (4)	4 (3)	3 (2)	IN GYM CLOTHES
9	HAND-TO-HAND COMBAT	1. A DEMONSTRATION OF TECHNIQUES WITH ELEMENTS OF SPEED (A SPECIALIST DOING THE GRADING) 2. PERFORMANCE OF A NUMBER OF CONTACT MOVES AND DEFENSES (FIGURE 6) 3. TWO-MINUTE SPARRING MATCH*				

* USING BOXING GLOVES IN USING BLOWS, THROWS, HURTING, AND CLOTHING TECHNIQUES

TABLE 1. SAMPLE STANDARD REQUIREMENTS IN SPECIAL PHYSICAL TRAINING

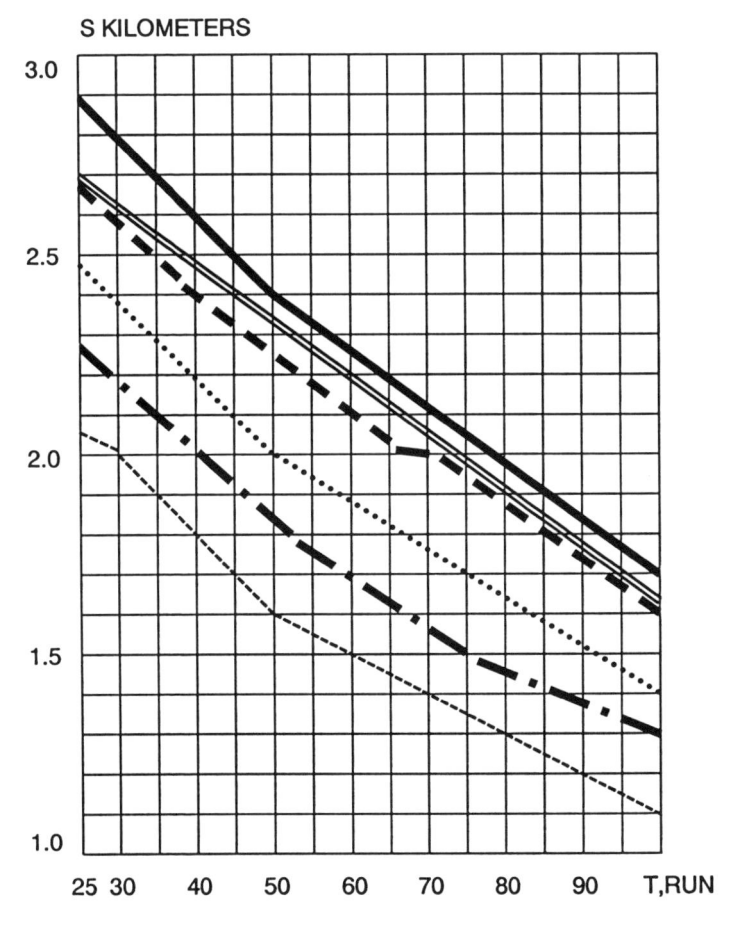

FIGURE 7. GRAPH OF THE NUMBER 1 RATINGS
(12-MINUTE, NO-BREAK RUN)

KEY:

	Men		Women
▬▬▬	Excellent	═══	Excellent
▬ ▬ ▬	Good	··········	Good
▬·▬·▬	Satisfactory	----------	Satisfactory

FIGURE 6. ELEMENTS OF ATTACK AND DEFENSE MOVES

hand-to-hand combat) can be included in one mixed exercise.

Such an approach furthers the all-around physical development of armed forces personnel, inasmuch as the physical exertion involves the major muscle groups, the cardiovascular and pulmonary systems, and lifts the emotional state of the participants. The result of this activity is increased [physical] solidness and capacity for work demands.

Going through the complex of training exercises makes easier a mastery of action skills for the trainees because of the frequent repetition of the skills in the course of training.

The individual course segments can develop into a complete training cycle. Uncomplicated assignments/exercises (eight to ten at most) can be performed uninterruptedly, one after the other (stream method) with a minor break of thirty to ninety seconds (work-interval method). This ensures continuous effect on all major muscle groups and a healthy stress for the cardiovascular and pulmonary systems. Especially beneficial exercises develop the physical qualities of speed, agility, endurance, and power.

Along with this, personnel who undergo this physical training should demonstrate that they are achieving results conforming to requirements by having to pass concurrent tests.

Tests in SPT are a regular result of the teaching process in given disciplines and of all work already performed. A test in a discipline can be developed from an array of intervals, that is, from tests in certain types of physical training, which present themselves in the course of training. A combined approach is allowed as a testing method, i.e., simultaneous testing in two, three, or more kinds of physical training.

This manual presents sample standards in SPT (Table 1) by which certification of physical qualities can be done. A general test in SPT can be put together from grading data from a whole period of instruction.

The most important physical qualities, with which trainees perform fundamental activities, are evidence of strength and general endurance.

Test Number 1

The level of development of a quality such as endurance may be determined by [this] test (twelve minutes, no break). To conduct it, there have to be a distance measured out in kilometers or

	EXAMINEE'S AGE	
PHYSICAL CONDITION	UP TO 31 YRS.	UP TO 36 YRS.
	NUMBER OF REPETITIONS	
EXCELLENT STRENGTH/ENDURANCE	4 (5)*	3 (4)*
GOOD	3 (4)	2 (3)
SATISFACTORY	2 (3)	1 (2)
UNSATISFACTORY	1 (2)	0 (1)

* GYM SUIT IS WORN

TABLE 2. RESULTS OF TESTS PERFORMED IN FIELD DRESS

FIGURE 8. CONTENT OF TEST NUMBER 2
A—EXERCISE 1 B—EXERCISE 2
C—EXERCISE 3 D—EXERCISE 4

FIGURE 8A. DO TEN PUSH-UPS

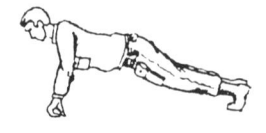

FIGURE 8B. MOVE FROM THE LEANING REST POSITION AND RETURN TO IT (SQUAT THRUSTS)

FIGURE 8C. LYING ON YOUR BACK, SWING YOUR LEGS TOGETHER BACK OVER YOUR HEAD, KEEPING YOUR HANDS BEHIND YOUR HEAD OR STRETCHED ALONG YOUR BODY, PALMS ON THE FLOOR. AFTER THE TENTH REPETITION, RETURN TO THE LEANING REST POSITION.

FIGURE 8D. JUMP UP FROM THE LEANING REST POSITION AND GO INTO A SQUAT, WITH ONE LEG SLIGHTLY IN FRONT OF THE OTHER AND YOUR HANDS LOCKED BEHIND YOUR HEAD. JUMP UP TO STAND ON TIPTOE, STRETCHING THE TORSO FULLY. RETURN TO THE SQUAT, EXCHANGING THE POSITION OF THE LEGS.

THE FOUNDATIONS OF SPECIAL PHYSICAL TRAINING

FIGURE 9. TEST NUMBER 3 EXERCISES

EXERCISE 9A. PERFORM OVERHAND PULL-UPS ON A CROSSBAR

EXERCISE 9B. JUMP UP FROM A SQUAT, HOLDING THE HANDS LOCKED BEHIND THE HEAD AND KEEPING ONE FOOT SLIGHTLY FORWARD OF THE OTHER. STRETCH THE TORSO TO THE MAXIMUM.

EXERCISE 9C. FROM THE LEANING REST POSITION, DO PUSH-UPS. THE CHEST MUST TOUCH THE FLOOR (AND THE ARMS MUST BE BENT FULLY, WITH THE BODY KEPT STRAIGHT). AFTER THESE EXERCISES ARE COMPLETED, A FIVE-TO-TEN-MINUTE BREAK CAN BE TAKEN.

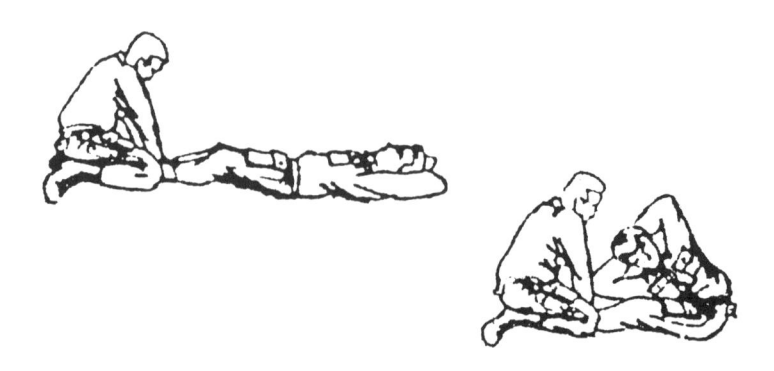

EXERCISE 9D. PERFORM SIT-UPS, KEEPING THE HANDS CLASPED BEHIND THE HEAD. IN EACH REPETITION, TOUCH THE RIGHT KNEE WITH THE LEFT ELBOW AND VICE VERSA. HAVE A PARTNER HOLD YOUR LEGS DOWN AT THE ANKLES. THE MAXIMUM NUMBER OF REPETITIONS POSSIBLE IS DONE IN TWO MINUTES.

EXERCISE 9E. ASSUME THE LEANING REST POSITION FROM THE SQUATTING POSITION, AND RETURN TO THE LEANING REST POSITION [SQUAT THRUSTS]. THE MAXIMUM NUMBER OF REPETITIONS IS DONE IN SIXTY SECONDS.

meters, a stopwatch, and test subjects.

The results in kilometers for a twelve-minute run are matched with data in a prepared table to determine a rating for development of endurance (excellent, good, or satisfactory). The runner should wear a track suit. No adjustments are made for bad weather.

Carrying out the test in a stadium is best, but it can be done in a forest, on a road, or at any other convenient place. The test doesn't take long, and it is simple and appropriate.

Figure 7 shows a determination of ratings graphically. The test allows self-testing with a hand-held stopwatch.

Test Number 2

The most important test to determine the level of strength/endurance development of service personnel, up to the age of thirty-six, is a series of tests composed of four types of general development exercises. Each exercise is done with ten repetitions by the instructor's count, with no breaks between exercises (ten + ten + ten + ten repetitions). The count is called at a medium pace. To start the test, the examinee assumes the leaning rest position.

This is a first series in a total complex of exercises.

To conduct the testing methodically, it is necessary to move immediately to the second performance of this complex of exercises. In so doing, several series have to be performed in succession. One instructor can calculate the results for a whole group of examinees. In the tests, one scorecard can be used for two examinees.

Test Number 3

Using this test's data, it is possible to determine quite accurately the level of physical strength of individuals in the thirty-one to thirty-six-year age group.

This test comprises five exercises (Figure 9). Test conditions are the performance of a maximum of repetitions with the highest level of technical accomplishment.

Examination results can be worked out with the aid of the arithmetical calculations of Table 3 that score each exercise and

KGB ALPHA TEAM TRAINING MANUAL

TYPES OF EXERCISES

SCORES	POINTS	OVERHAND PULL-UPS		JUMP FORM SQUATTING POSITION		PUSH-UPS		SIT-UPS		SQUAT THRUSTS	
1	2	3		4		5		6		7	
EXCELLENT	100	100*	18**	100	95	100	60	100	85	100	41
		97	17	98	94	98	59	98	84	96	40
		94	16	96	93	96	58	96	83	93	39
		90	15	94	92	94	57	94	82	90	38
		86	14	92	91	90	55	92	81	85	37
		82	13	90	90	88	54	90	80	80	36
		78	12	88	89	86	53	88	79	75	35
		75		85	87	84	52	86	78		
				84	86	82	51	84	77		
				83	85	80	50	82	76		
				80	82	78	49	80	75		
				75	77	75	48	78	74		
								76	73		
	75							75	72		
GOOD	74	74	11	74	76	74	47	74	71	74	34
		70	10	73	75	70	45	73	70	67	34
		65	9	72	74	68	44	72	69	62	32
		60	8	71	73	66	43	71	68	59	31
		55	7	70	72	65	42	68	65	55	30
		50	6	68	70	63	40	63	60	50	29
				63	58	35	58	55			
				58	53	30	53	50			
				53	52	29	52	49			
				51	50	27	50	47			
	50			50	52						

* POINTS
** NUMBER OF REPETITIONS

TABLE 3. SCORECARD FOR TEST NUMBER 3

1	2	3		4		5		6		7	
SATISFACTORY	49 ↓ 24	49 40 39 33 32 35	5 4 3 29 28 25	49 43 40 38 37 37 36 35 34 30 29 26	51 45 42 40 39 35 37 36 35 28 25	49 48 43 41 40 39 38 36 30 11 10	26 25 20 19 18 17 16 15 12	49 43 38 33 28 25	46 40 35 30 25 22	47 43 40 36 32 28 25	28 27 26 25 24 23 22
UNSATISFACTORY	24 ↓ 1	24		24 19 14 9 4 3 2 1	25 20 15 10 5 4 3 2	24 20 16 12 10 8 1	9 7 5 3 2 1	24 18 13 8 6 4 1	21 15 10 5 3 1	22 18 15 13 9 5 2 1	21 20 19 18 17 16 15 14

SUMMARIZED SCORING INDEX (POINTS)
EXCELLENT: 500-375
GOOD: 370-250
SATISFACTORY: 245-125
UNSATISFACTORY: 120-18

excellent score equals an ideal of 500 points (upper limit) down to 375 points (lower limit: 75 x 5).

NOTES

[1] In this exercise the chest has to touch the ground.
[2] The knees are kept straight in the thrust.
[3] V.I. Sukhotskii. *Physical Training in the U.S. Army*. Moscow: Voenizdat, 1975.

CHAPTER 2
MOVEMENT; OVERCOMING OBSTACLES; PENETRATING/STORMING BUILDINGS

The basic methods of movement in a group or individually are walking, running, movement by foot or on skis, various runs, cross-country runs, and forced marches. All these are the most accessible ways to achieve general physical fitness. Regular morning exercises are a great help in this. The exercises themselves are twenty-minute runs, alternating with ten minutes of such simple exercises as push-ups, pull-ups, squat jumps, and lifting dumbbells or weights. These can be done at any time of year, in any weather, no fewer than three times a week, with clothing suitable for the weather.

Three-kilometer runs are recommended for training units. At the end of a month's training, a forced march of six kilometers should be organized for the units.

An excellent means of developing comprehensive fitness is a long-distance run every two months, with individual scoring. The runs should be done along routes suitable for such; observation of performance can be done by mobile scorers.

Skiing training strengthens the health, hardens the physique, and effectively increases physical fitness.

The instructional activity, in its preparatory, beginning, and concluding segments, constitutes the basic form of physical training. The goal of each segment is different, but in the process of completing each one, trainees must receive psychological conditioning.

The instructor in charge of each activity defines the concrete goals and aims of the body of the work plan and a methodical realization of the plan.

An essential goal in teaching basic movement methods is the development of physical qualities in combination with psychological conditioning, utilization of acquired skills, and knowledge of action in operations.

Movement under combat conditions means remembering the basic rule of appearing where the enemy doesn't expect you.

Any kind of movement demands:
- Not being seen in a given place
- Moving quickly but as silently as possible
- Observing and noting everything while moving and
- Being ready for the unexpected

Natural barriers such as streams, ditches, wooded heights, and minor landforms can be gotten past without special methods.

To penetrate or bypass man-made obstacles and positions, previous training is needed. The training is based on experience of the latest-known requirements, which address movement by groups that have learned advanced techniques.

GENERAL SYSTEMATIC DIRECTIONS FOR TEACHING BASIC METHODS OF MOVEMENT

Getting past different obstacles and positions raises moral-psychological hardiness and should have a place in all planned training activities. It is recommended that, in the course of a training segment in SPT, five to seven minutes be allotted for warm-up exercises when using the stream method to learn the overcoming of two or three simple obstacles. In dry or very cold weather, it makes sense to practice on low obstacles to safeguard against falls or accidents. Methods used must foster success. Without safeguards more complex obstacles will not be overcome, especially by an individual. To avoid injuries, the trainer or instructor must maintain discipline, using authority

MOVEMENT; OVERCOMING OBSTACLES; PENETRATING/STORMING BUILDINGS

and presence. The instructor himself has to demonstrate the most complex obstacle penetration techniques, observing safety and self-protection rules. Methodically given instructions mean that the instructor will be understood.

To teach movement methods, it is best to choose wooded terrain with steep slopes. Learning various methods (e.g., crawling or short rushes) can take place during training for cross-country runs—first without carrying weapons and later with weapons and full equipment, day or night.

FUNDAMENTALS OF MOVEMENT AND OVERCOMING OBSTACLES

Walking in a crouch can be used in high grass, grain fields, or brushy areas—that is, when it is imperative to stay hidden from the enemy. In approaching the enemy, one moves silently, with slightly bent knees, stepping in a heel-to-toe fashion. To move faster, one runs. In military practice, soldiers use a uniformly long stride, swift running, or running alternated with walking and crawling to overcome various obstacles. Running is done lightly, loosely, and without excessive straining. Soldiers leap during movement to get across holes, ditches, incidental objecs, and to jump down into cover. The best method for concealed movement is the low crawl, with a target: the

FIGURE 10. LOW CRAWL UNDER A TREE

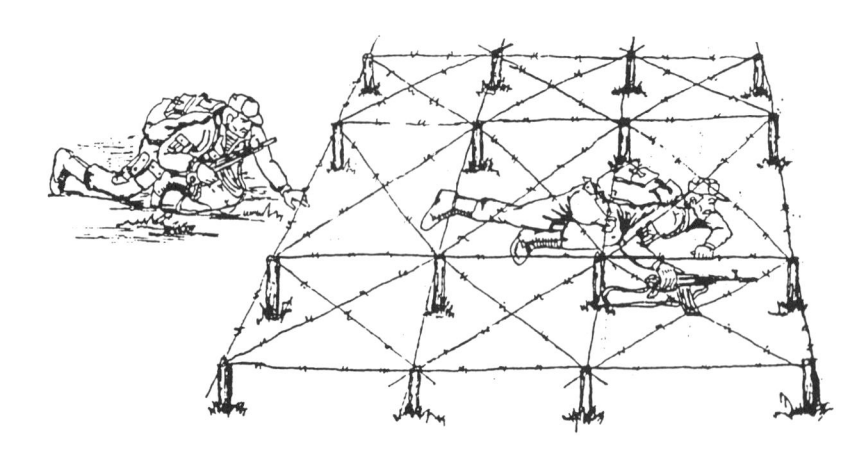

FIGURE 11. LOW CRAWL UNDER BARBED WIRE

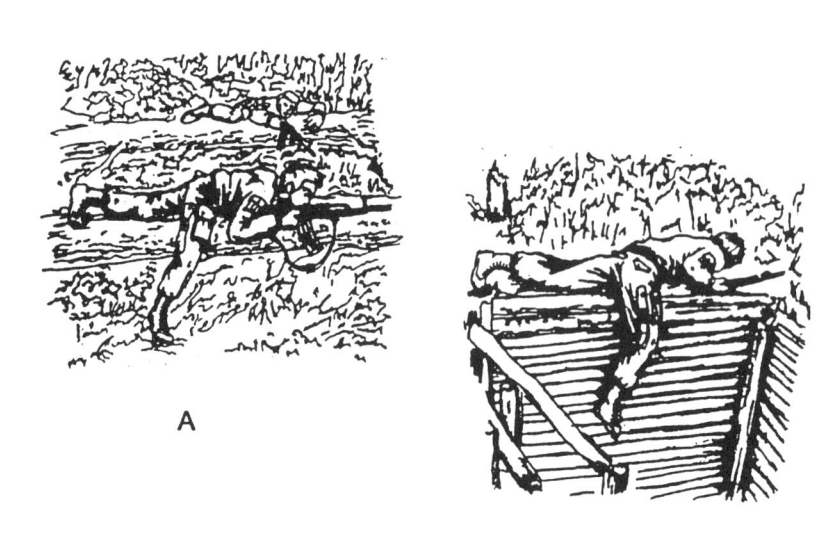

A

B

FIGURE 12. CLAMBERING OVER AN OBSTACLE
A—OVER A TREE
B—OVER A REINFORCED POSITION

MOVEMENT; OVERCOMING OBSTACLES; PENETRATING/STORMING BUILDINGS

FIGURE 13. GETTING OVER BARBED WIRE
A—WITH A LADDER
B—WITH A MAT

FIGURE 14. CREEPING ACROSS BARBED WIRE, USING A PLANK AND CLOTHING

concealed approach to capture or kill an enemy. The low crawl is done hugging the ground, using its irregularities, alternately moving the arms and legs, moving forward with the weapon in the right hand (Figures 10 and 11). In this, it is essential to keep the muzzle slightly ahead of oneself and above the ground's surface, snow, water, or dirt.

Individual natural and man-made obstacles and positions can be overcome by crawling and clambering, utilizing various devices when needed. Such methods are show in Figures 12, 13, and 14.

In both circumstances (Figure 12), one has to lie on the obstacle and move sideways, keeping the face in the direction of movement.

Some obstacles are cleared by using a hand grenade: for instance, to move quickly through a barbed wire entanglement.

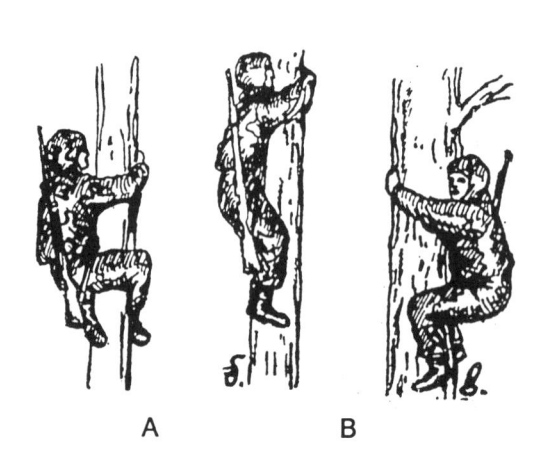

FIGURE 15. CLIMBING A TREE
A—GRIPPING WITH THE LEGS
B—STRAIGHTENING THE LEGS

FIGURE 16. MUTUAL HELP IN CLIMBING

MOVEMENT; OVERCOMING OBSTACLES; PENETRATING/STORMING BUILDINGS

Certain points are gotten past or into by climbing, which is used for storming occupied positions, buildings, or forests, elevations, and so forth (Figures 15, 16, and 17).

Using a horizontally rigged rope, as show in Figure 18, you

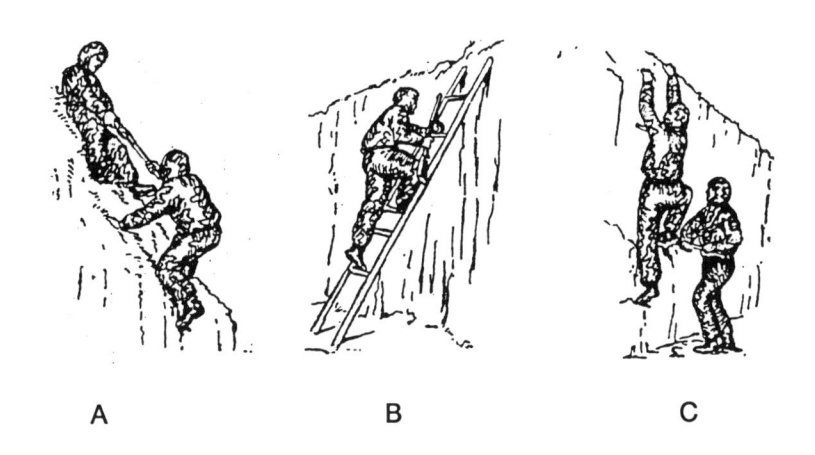

FIGURE 17. AIDS IN SURMOUNTING AN OBSTACLE
A—UP A STEEP SLOPE WITH A COMRADE HELPING WITH AN ENTRENCHING TOOL
B, C—UP A BARRIER WITH A LADDER'S OR COMRADE'S HELP

FIGURE 18. CROSSING A RAVINE BY ROPE

can get across ravines, water courses, gullies, and mountain streams. A very agile and strong person should fasten or hold the rope tightly.

MOVEMENT UNDER SPECIAL CONDITIONS

For groups engaged in storming an objective, it may come to creeping over roofs; negotiating entrenchments, ditches, and depressions; working through brush or marshes; wading silently; and so on. See Figures 19 and 20 for a general depiction of these movements.

FIGURE 19. MOVING ON ROOFS
A—FACING THE ROOF
B—ON ALL FOURS
C—BACK TO THE ROOF

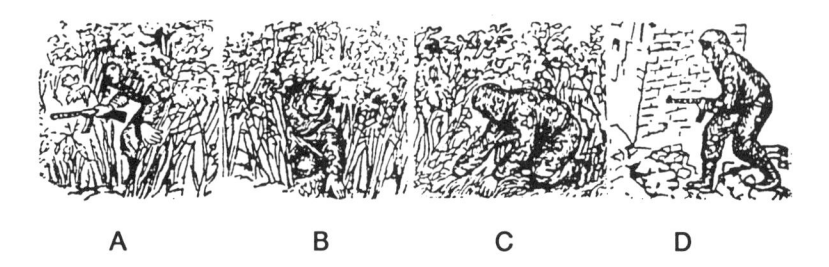

FIGURE 20. MOVEMENT IN VARIOUS CIRCUMSTANCES
A—IN THE BRUSH
B, C—IN MARSHES
D—THRUGH RUBBLE ALONG A WALL

MOVEMENT; OVERCOMING OBSTACLES; PENETRATING/STORMING BUILDINGS

Movement in general, and particularly in special circumstances, demands carefully following the rules of concealment, avoiding in any way possible the group's being discovered. When using roads, avoid standing out against the background. In small towns, stay close to walls or fences; move in from woods or gardens. In forests, stay in the shadows. On open ground, make use of irregularities of the terrain. In bad weather, move quickly to take advantage of the noise made by winds, rain, and so on.

SPECIAL FEATURES OF NIGHT MOVEMENT

Night movement has its own special characteristics and differs from that at other times of day. It comprises standing upright, crouching, going on all fours, and crawling. Moving quietly at night involves feeling out the ground with the toe before standing firmly. During such movement, the hand not holding the weapon should be kept straight out in front, in the

FIGURE 21. NIGHT MOVEMENT, STANDING UPRIGHT

A   B

**FIGURE 22. NIGHT MOVEMENT IN A LOW CROUCH
A—FACING THE FRONT
B—FACING TO THE SIDE**

direction of movement (Figure 21).

Moving in the dark demands being cautious, feeling out the ground ahead, and not disturbing objects in the way. In running at night, steps should be short with the feet lifted just a bit, and the left hand should be held straight out in front. Night movement in a crouch is shown in Figure 22.

When moving, it is absolutely necessary to keep your weapon against the chest or in the right hand. Moving forward should be done with one side, usually the left, turned somewhat in the direction of movement. You should crouch to stay concealed. Lean the body forward slightly and keep the knees bent (Figure 23).

Movement has to be silent, "in stealth" (Figure 24).

FIGURE 23. MOVING IN A CROUCH

FIGURE 24. MOVEMENT "IN STEALTH"

When nearing the enemy, move cautiously. Take small steps, placing the feet firmly, keeping your balance, and stopping often. The halts are needed to "listen" to the enemy because during movement you hear badly, and the enemy can pick up the sounds of someone approaching.

LEAPING NATURAL OBSTACLES

The need to get past such natural obstacles as channels, ditches, depressions, and so on, can come up in the course of movement.

MOVEMENT; OVERCOMING OBSTACLES;
PENETRATING/STORMING BUILDINGS

A depression in the ground can be jumped by launching yourself from the edge, using either one foot or both feet (Figures 25 and 26); the jumps can be done from a running or standing start.

Obstacles can be overcome one after the other while moving,

FIGURE 25. LEAPING FROM THE EDGE

FIGURE 26. DYNAMIC LEAP FROM A STANDING START

without [group members] bumping into one another, and while keeping each other from unpleasant landings and falls. Jumps using one hand can be done to clear earth banks, fallen trees, low fences, and other barriers of no great height (Figure 27). The free hand should help guard against slips and falls.

Long jumps down from walls or fences are shown in Figures 28 and 29. Hold your weapon in your hands or sling it across your chest or over your back. If your rifle is slung over your

FIGURE 27. JUMPING A FALLEN TREE, USING ONE HAND

back, after landing (with knees bent), use your free hands to help a comrade. You can also use ladders, boards, logs, ropes, and so forth in this.

RUNNING AND CRAWLING

In combat, running is employed in all circumstances and in any locale. It should be alternated with short halts, resting on one knee (Figure 30) and lying prone (Figure 21)—with your weapon ready to fire.

The way to do this is to first stop and drop to one knee; then, leaning on the extended left arm, lie down and stay ready to fire from the

FIGURE 28 JUMPING DOWN FROM AN OBSTACLE

MOVEMENT; OVERCOMING OBSTACLES; PENETRATING/STORMING BUILDINGS

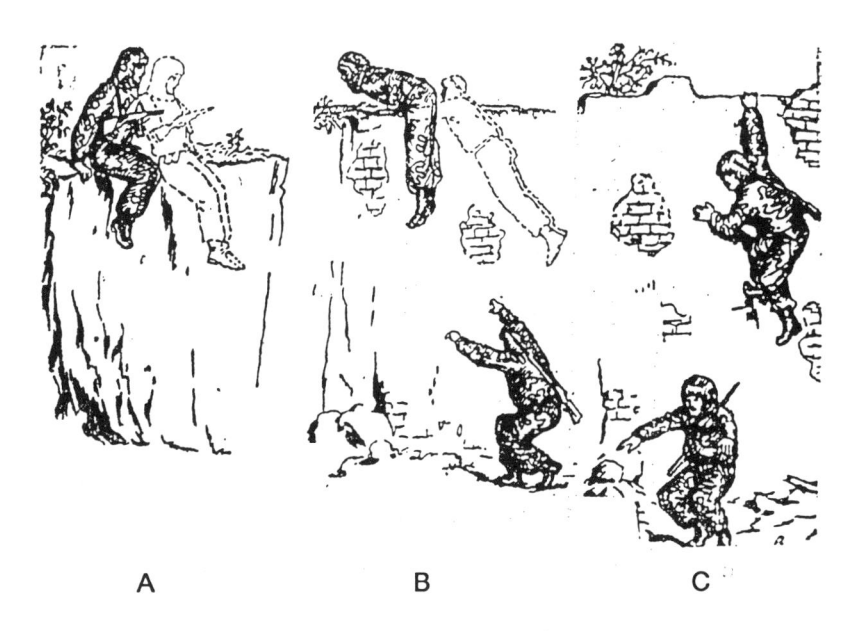

FIGURE 29. JUMPING DOWN FROM AN OBSTACLE
A—FACING FORWARD
B—FACING BACKWARD
C—SIDEWAYS

FIGURE 30. RESTING ON ONE KNEE

FIGURE 31. LYING PRONE

prone position. You can go from running to the prone position, using the technique of "reeling in," that is, by pulling yourself forward with the extended free arm and hand (Figure 32).

To perform this maneuver with consistency, the following needs to be done:

KGB ALPHA TEAM TRAINING MANUAL

FIGURE 32. SWITCHING FROM RUNNING TO
THE PRONE POSITION
A—RESTING
B—LEANING ON THE EXTENDED ARM
C—FINAL POSITION

• Rest, with a turn to face right
• Come down on your left side, using the left arm and hand
• Change to the final, prone position, with your weapon ready for firing

Advancing by crawling, weapon in hand, is used for

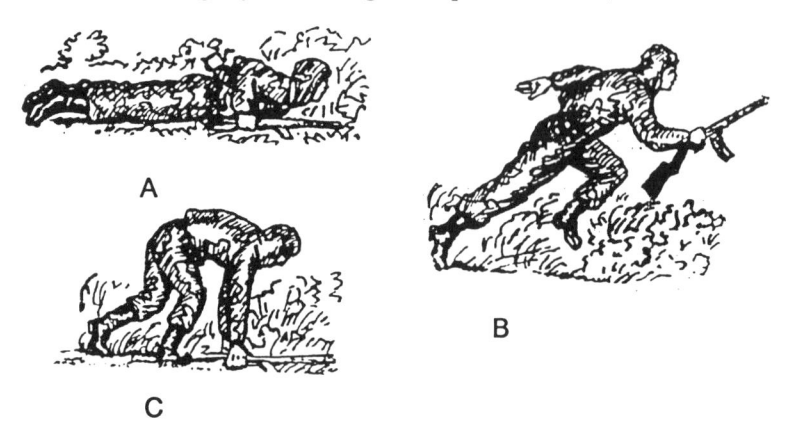

FIGURE 33. SWITCHING FROM THE PRONE TO THE
STANDING POSITION TO RUN
A—AT FULL PRONE
B—RISING
C—MOVING

MOVEMENT; OVERCOMING OBSTACLES; PENETRATING/STORMING BUILDINGS

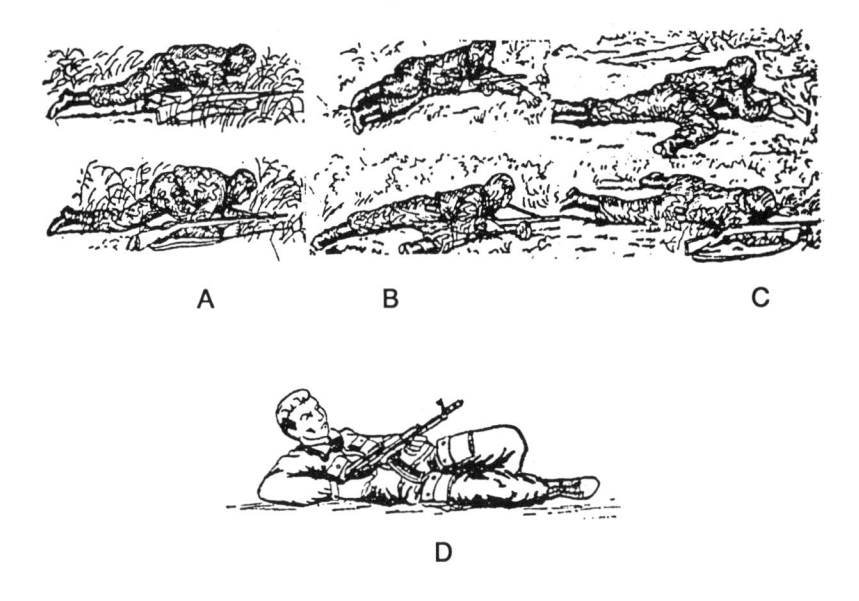

FIGURE 34. WAYS OF CRAWLING
A—ON THER KNEES C—LOW CRAWL
B—ON THE SIDE D—ON THE BACK

concealed approaches, to take out a sentry, to get close to an objective, to cross fields of fire, and for other situations.

Employment of the different methods of movement depends on locale, terrain and vegetation, and other conditions. It is very important that training in concealed movement be combined with quick jumps to the feet; short, swift dashes; and subsequently hitting the ground for crawling (Figure 33).

The ways of crawling, positioning the weapon, and working the arms and legs are shown in Figure 34.

Using these methods, you can move quite fast. The low crawl is the quietest compared to the other methods, but it is notably slower and uses up a lot of energy.

Stealthy, creeping advances can be made up of combinations of methods: crawling, jumping to your feet, running, falling, crawling, and so on. Whenever you are carrying a weapon and performing these, take special care to avoid injuries. During the teaching of crawling methods, trainees should wear clothing appropriate to the season in order to maximize dexterity and agility, as well as to develop skill in and knowledge of action under demanding conditions.

MOVEMENT IN MOUNTAINS

Operating in mountains requires such qualities as fitness, caution, agility, patience, close observation of hostile forces and regions, and not losing your orientation. Trainees absolutely have to master the fundamentals of moving in mountains and conquering alpine barriers, the most important of which are crags and cliffs. Mountain training can be included in all physical training activity.

A range of factors influences movement in wooded mountainous areas, and they have to be factored into alpine combat operations. The difficulty of keeping one's bearings arises from the lack of vegetation and normal points of orientation, the results of unexpected changes in weather, and reduced visibility in fog, low clouds, and rain. There is a limited number of roads and paths of any kind. Movement is also disrupted by falling rocks, avalanches, drifts, and slides. Add to these factors that an enemy can easily discover you through a fire's being lit. Carrying only what is needed to carry out the mission is a part of movement in mountains. To avoid danger, it is imperative to know and use precautions and self-protection.

A most important condition of successful alpinism, to maintain safety and overcome difficult, physical features, is *not* to carry any unnecessary thing in your hands. The hands must be left free as much as possible. Weapons should be slung around the body or across the back in such a way that they do not hinder movement, but are always ready for use en route.

In dangerous spots, a group of three or four individuals should move connected to each other by rope (Figure 35).

Each one in the group wraps a part of the rope around his chest, securing the loop in a knot, as shown in Figure 36. To keep the loop from slipping off the chest, tighten your part with

MOVEMENT; OVERCOMING OBSTACLES; PENETRATING/STORMING BUILDINGS

FIGURE 35. MOVING CONNECTED BY ROPE

an extra length, passing one end under the loop in back, then past the neck to put the other end under the loop in front. Tying the ends together is essential.

To keep the feet from slipping on slopes, wrap your footgear with rope or wire (Figure 37).

To avoid falls and accidents when moving in a group on one rope, it is necessary to choose suitable projections, ground surfaces, and depressions that are stable and can be used for

FIGURE 36. SECURING THE ROPE; INDIVIDUAL WEAPONS IN A SLING

FIGURE 37. WAYS OF WRAPPING FOOTGEAR

FIGURE 38.
DESCENT
USING AN
ENTRENCHING
TOOL

FIGURE 39.
DESCENT, USING
A ROPE AND
TREE

FIGURE 40. USE A
ROPE ACROSS
THE SMALL OF
THE BACK, WITH
FEET PROPPED

support or additional help in movement. To ensure smooth movement, use rifle butts, entrenching tools, or sticks with sharpened ends, keeping them ready for use in descents or traversing dangerous spots (Figure 38).

For safety, carefully secure a rope around a tree, over a shoulder, or across the small of the back (Figures 39 and 40).

Going up or across slopes calls for methods that utilize a rope secured by a comrade, or going on all fours, holding onto projections, shrubs, or grass for help (Figures 41 and 42).

Figures 38 through 40 show the simplest ways of descending steep slopes.

On rock-strewn slopes, place your feet carefully, watching out for shifting footing. Figure 43 shows the basic method of ascending a grassy and rocky slope.

MOVEMENT; OVERCOMING OBSTACLES; PENETRATING/STORMING BUILDINGS

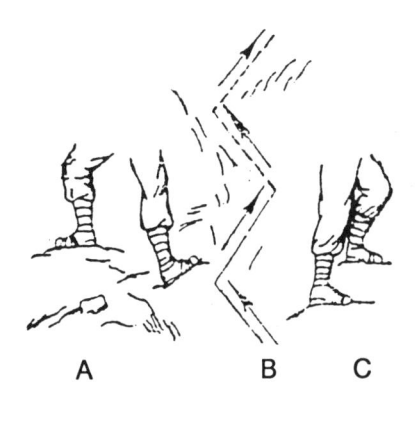

FIGURE 41. UPRIGHT ASCENT
A—STRADDLING
B—ZIG ZAG
C—BY STEPS

FIGURE 42. ASCENT ON ALL FOURS

FIGURE 43. ASCENT FACING FORWARD

MOVEMENT IN DESERTS

Sandy areas abound in dunes. Regions covered by sand are characterized by high average temperatures and a lack of water. In such regions low population density, an insufficiency of watered areas, and strong winds destroy your orientation. Movement by foot is slowed by half, and you have to put your feet on spots where the sand is firmest. The best times for travel are at night and in the morning before sunrise. During the day, a light head covering the color of the sand should be worn to prevent sunstroke. Salt tablets are best for getting through deserts. When crawling, you have to cover your weapon muzzle somehow. You must not get separated from the group. Expend your energy sparingly.

In addition, you should not forget that open spaces provide excellent visibility, which makes it easy to aim fire from a distance and hard to organize natural concealment. In arid and semiarid regions, an enemy will most quickly set up defense of populated spots, oases, wells, crossroads, and crests of elevations and mounds.

OVERCOMING MAN-MADE OBSTACLES AND POSITIONS

To establish a foundation for learning movement against and overcoming man-made obstacles of various construction, a sports training facility must be used to develop the qualities of strength, dexterity, and agility. Methods to get through obstacles are worked out through experience gained in the course of practical exercises. Given that overcoming individual obstacles carries a certain element of risk, studying (by reading and observing) to become familiar with standard methods is essential.

The Springboard for Landing and Jumping

This is used for ground-level preparation and physique development (along with psychological conditioning) for trainees before they perform parachute jumps. Trainees climb onto it, using their hands. Figure 44 shows jumps performed from various heights and a landing on both feet with the knees bent. In winter, landing should be on snow; in summer, it should be onto sand or sawdust.

High and Low Bars

These are for developing arm strength. The methods used are to hang by the hands (high bar) or to hold yourself up on extended arms (low bars) and move along the bars (Figure 45).

Peaked Ladder

This is used to develop trainees' agility and strength. The method is to get over a peak without using your hands, to the

MOVEMENT; OVERCOMING OBSTACLES;
PENETRATING/STORMING BUILDINGS

FIGURE 44. JUMPS FROM A SPRINGBOARD

FIGURE 45. MOVING ALONG BARS
A—HIGH BAR
B—LOW BARS

next slope of the ladder, using your hands to climb it, but moving and staying upright with only your legs whenever possible (Figure 46).

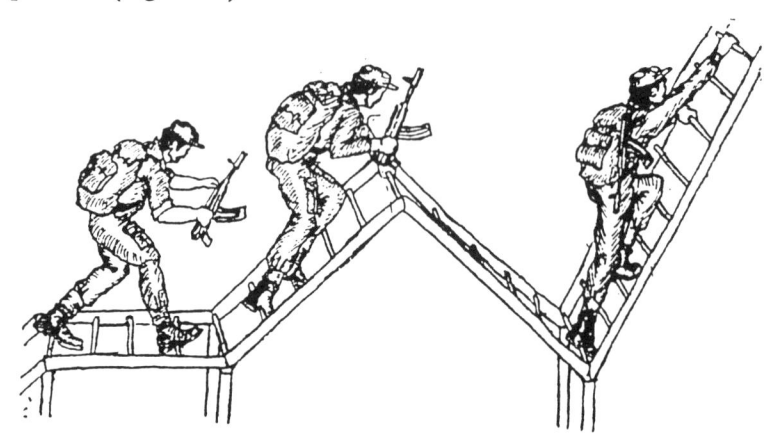

FIGURE 46. MOVEMENT ON THE PEAKED LADDER

Rolling Log

Performing movement on this is an elaboration of knowing how to use forest materials to cross deep ravines, streams, gullies, and other depressions. When carrying any extra weight, for safety and to avoid falling from a great height, you should straddle the log, staying upright and propping yourself up with your hands. If you lose your balance, then it's under the log, progressing by moving your gripping hands and legs along the log in alternation (using a "hug"), as when climbing a rope (Figure 47).

High Plank

Negotiating this strengthens, above all, a trainee's spirit, and also engenders a knowledge of how to move in heights without fear. Safety devices, such as a cable or a carabiner on the cable, are used (Figure 48).

Climbing up onto a high plank can be done by vertical or leaning ladder. Use a ladder to get off the plank. Exercises on the plank are done only with safety devices in use.

MOVEMENT; OVERCOMING OBSTACLES; PENETRATING/STORMING BUILDINGS

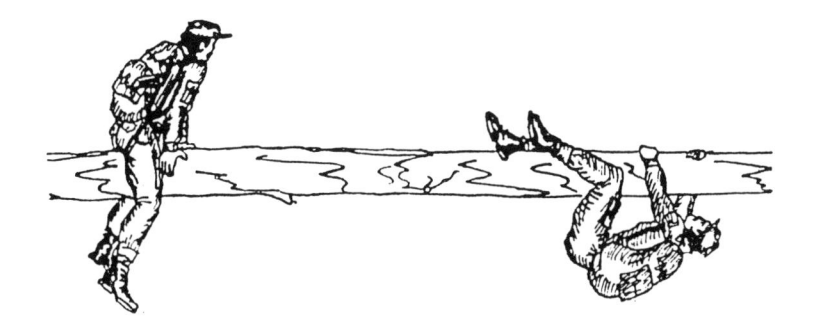

FIGURE 47. GETTING ACROSS ON A ROLLING LOG

FIGURE 48. GETTING ACROSS ON A HIGH PLANK

Pole or Rope

Climbing these develops agility as well as power in the hand and arm muscles. The method is shown in Figure 49.

This is especially valuable in climbing heights. On the command "one," grasp the pole (or rope) as high up as possible, grip the pole with your feet, with the lower foot's instep against the pole and the upper foot resting on the lower—at the same time gripping the pole with your knees.

On "two," straighten your legs at the knees as if to stretch to your full height, supporting yourself with your gripping feet.

On "three," slide your hands smoothly up the pole and then tighten their grip. Repeat the foot and leg movements, as in the first step, and continue on up.

Solid Fence Obstacle

Dealing with this obstacle is done by pulling yourself up by the hands, and either using a foot or rolling sideways on your stomach to get over (Figure 50). In learning this one, a comrade can help.

"Two-Tier" (or More) Barbed Wire Entanglement

This one has to be approached creatively. Under combat conditions, any materials can be used as aids, e.g., boards, logs, and/or clothing. A board can be set up as a ramp against one side of the entanglement, and you can mount to it that way. On top, clothing can be thrown onto the entanglement, and, once you stand on the clothing, you can draw the board up to be used as a way over the wire or as a ramp down off another side (Figure 51).

FIGURE 49. POLE/ROPE CLIMBING

Conduit Pipe of Different Diameters

For invading inhabited buildings, it is a must to learn how to navigate underground communication routes where you are faced with constricted movement: crawling and using your elbows (Figure 52). A pack is not carried on the back; it is carried in one hand or pushed ahead of you through tight-diameter pipes.

Storming a Building, Using Door Frames and Window Sills

Training in overcoming such obstacles develops skills in

MOVEMENT; OVERCOMING OBSTACLES;
PENETRATING/STORMING BUILDINGS

A B

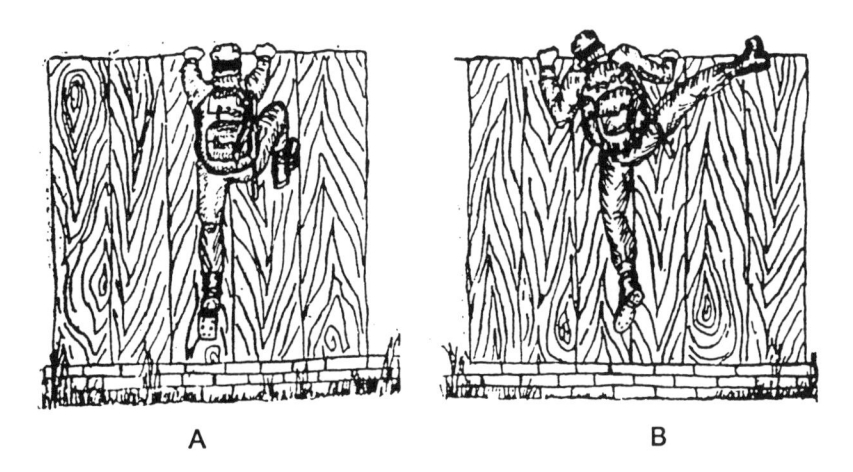

FIGURE 50. CLEARING A SOLID FENCE
A—GAINING A HOLD
B—TRAVERSING THE BARRIER

FIGURE 51. OVERCOMING A BARBED WIRE
ENTANGLEMENT

FIGURE 52. GETTING THROUGH CONDUIT PIPES
A—LARGE DIAMETER
B—SMALL DIAMETER

maneuvering in inhabited areas. These structures are not taken by a single person acting alone, but by small teams that provide mutual support internally. (In training for this, individuals of great strength and agility should be supplied with poles, ladders, and ropes with grappling hooks.) However, the basis of collective action lies in the training of each individual. In studying such obstacles, trainees should start with small pieces of information and gradually increase their mastery of methods of penetrating buildings by scaling them. In the first stage safety is imperative. Ropes are to be used more than in actual operations, ladders are to be kept at hand, and help is given in ascents and descents. Ascents and traverses are shown in Figure 53.

When operating in pairs, one team member has to follow the other closely, with each ready to render help to the other. In summer, there should not be more than two two-man teams on a wall at the same time; in winter there should be only one team. Only rubber-soled boots should be worn for this activity.

Other obstacles, e.g., spars or wire cables, do not require much training beforehand because mastering them is quite easy. In practicing, troops can act confidently on their own

MOVEMENT; OVERCOMING OBSTACLES; PENETRATING/STORMING BUILDINGS

FIGURE 53. SCALING A WALL VIA DOOR AND WINDOW FRAMES

immediately after the routines are explained, without jeopardizing their safety or taking serious falls. These obstacles are of minimal height, which guarantees safety.

Jumping from Heights

Methods are shown in Figure 54. These are worked out under field conditions, and can be further elaborated in training.

It should be noted that such jumps are made from a standing start and are, therefore, relatively simple to do.

It is significantly more difficult to jump from a moving object. In combat, decisions often have to be made in an instant; therefore, troops have to leave their motor transport and, landing

A   B

FIGURE 54. JUMPING FROM HEIGHTS
A—INTO A GULLY
B—FROM A ROOF

safely, go into battle to secure victory, to win. Figure 55 illustrates techniques for jumping out of moving vehicles. This difficult action is performed only after considerable training previously, which includes the following: tumbling forward and backward over objects on the ground and when running; somersaults and leaps into and off of mats moved at different

MOVEMENT; OVERCOMING OBSTACLES; PENETRATING/STORMING BUILDINGS

speeds, and so on. After this comes jumping from slow-moving vehicles, first into deep snow, then onto winter road tracks, and finally onto solid ground. Along with this goes practice in tumbling forward and sideways.

These exercises are begun without any weapons carried. Then, according to how well the skills are learned, trainees move on to practicing the exercises with weapons carried. (At first, wooden replicas wrapped in strips of cloth can be used. For final exercises, training and regular rounds—blank and live—can be carried.)

Consistency in this training has to be strictly maintained—whether in gymnasiums, in streets, or on town squares, or when on the move. In all instances, special attention has to be paid to conditioning the back, the vertebrae, and adjoining areas of the body.

In going through the actual jumping, vehicle speeds should vary from five to ten kilometers per hour. At increased speeds (greater than twenty kilometers per hour), exiting should take the following form: hanging by hand from the back of the vehicle, feet touching and running along the ground for a short distance, gradually letting go with one hand, then the second one.

Until balancing skills are acquired for such quick exiting from moving vehicles, trainees are categorically forbidden to jump out of faster-moving auto transport. With the goal of minimizing injuries and maximizing good performance of this action, intensive warm-ups are a must. Personnel must wear bulletproof vests and helmets.

Getting aboard a truck while it is moving is not an entirely easy maneuver. There are a few ways; two of the most effective are shown in Figures 55b and c.

To train for this, the following is recommended: jumping over a gym horse, and over a trestle; jumping off pedestals (of varying heights), onto a high pedestal from a run, with the help of a comrade, onto a moving pedestal, and so forth.

To get into a vehicle skillfully, good coordination and help from comrades is necessary, thereby ensuring safety for all. It is a must that, in the beginning, speeds are kept low on hard-surface roads, only gradually raising the speeds on increasingly

KGB ALPHA TEAM TRAINING MANUAL

A

B

FIGURE 55. GETTING IN AND OUT OF AUTO TRANSPORT
A—JUMPING OUT
B—JUMPING IN (1ST METHOD)
C—JUMPING IN (2ND METHOD)

rough road surfaces (e.g., wet surfaces or deep snow).

After they have been mastered, all the techniques described here must be constantly practiced with moving vehicles, taking every precaution and enhancing and refining all aspects.

CROSSING WATER BARRIERS

Getting across bodies of water, especially when using material objects to help, is an important part of combat training. For success in this, you have to know how to move in water: to swim, to dive, to skillfully use devices to help, to become tough, to not invite hypothermia. Best of all is to choose a limited part of a stream and ford it. No matter how this is arranged, whether the ford is guarded from or is part of a large stream, there should be no physical aids or closer shore-to-shore distances: fording should demand swimming.

Fording by swimming should be done with a complete absence of objects that can increase buoyancy. Before fording, footgear should be removed, tightly fastened pants loosened, pockets emptied, sleeve and collar buttons undone, and footgear tucked under the belt so that they don't sink in the water. Packs or any other carriers should be compressed to the limit and their openings secured tightly. The personal weapon goes on the back (Figure 56). Individuals who are strong swimmers can proceed without removing their boots.

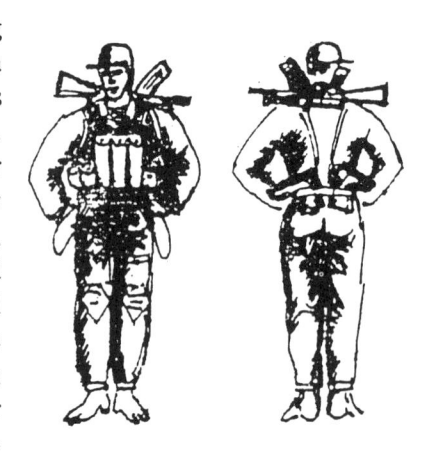

FIGURE 56. PREPARATION FOR FORDING

Fording by swimming is done without lifting the arms or taking a stroke.

In contrast, when fording by swimming and using different physical objects as aids to flotation, look for those quickly and easily found in the local inventory of equipment. You can experiment with plastic bags (big and small), football bladders, inflatable toys, light canisters, canteens and other hollow items, and even pieces of Styrofoam and foam rubber. Before swimming, all clothing must be removed, put in backpacks along with pairs of footballs or other inflatable balls (not completely filled with air) (Figure 57).

A tightly closed-up pack is placed in a large plastic sack (with no holes), which is tied tightly. Everything can be placed in such a sack, and the sack can serve then as a pack, with only it getting wet and the contents staying dry. Figure 58 shows a pack's contents in a plastic sack. Using it as a float, with a weapon resting on top is shown in Figure 59.

Once the sack is filled with gear and flotation devices, it can

MOVEMENT; OVERCOMING OBSTACLES;
PENETRATING/STORMING BUILDINGS

FIGURE 57. AIDS TO FORDING

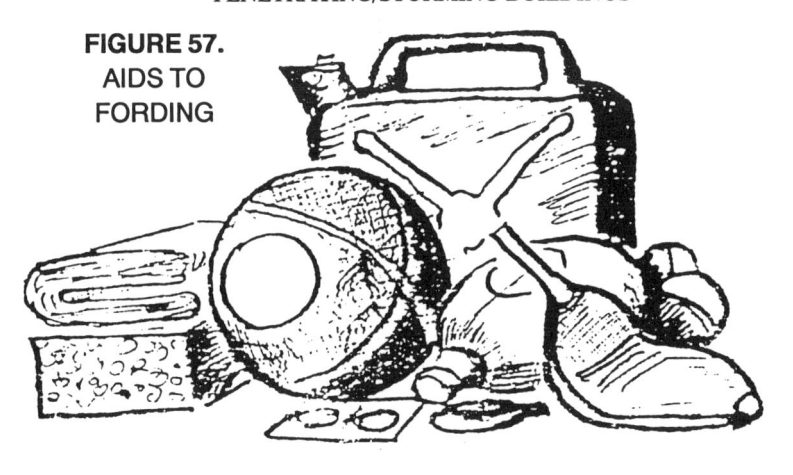

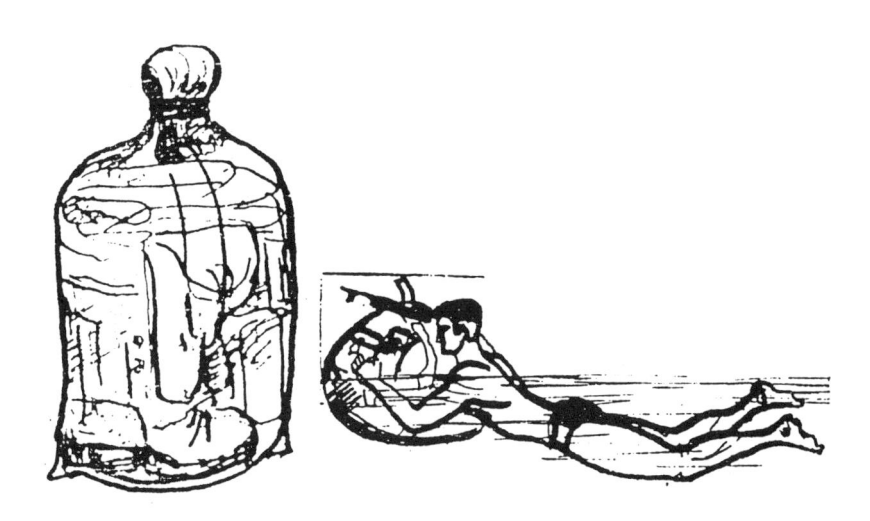

FIGURE 58. PACKED PLASTIC SACK

FIGURE 59. USING THE PLASTIC PACK IN FORDING

be secured to your shoulders with straps, as on a field march, leaving the arms and legs free for swimming (Figure 60). In this way, the pack-"raft" takes care of a big part of the swimming and the transport.

Soviet army doctrine recommends the use of issue rain capes and shirts, which can be stuffed with hay, straw, brushwood, reeds, or other buoyant material (Figure 61). Naturally, you cannot carry these to wherever, and they are not found in the steppes or in arid regions—and if they exist, it takes a lot of time and a knack in gathering them for fording use.

FIGURE 60. FORDING/SWIMMING WITH THE PLASTIC PACK

A weak swimmer can tie two logs (or filled plastic sacks) together to get across a water body. They are placed under the arms, so as to carry a person. Such a raft is shown in Figure 62.

Figure 63. Rope-Aided Temporary Ford

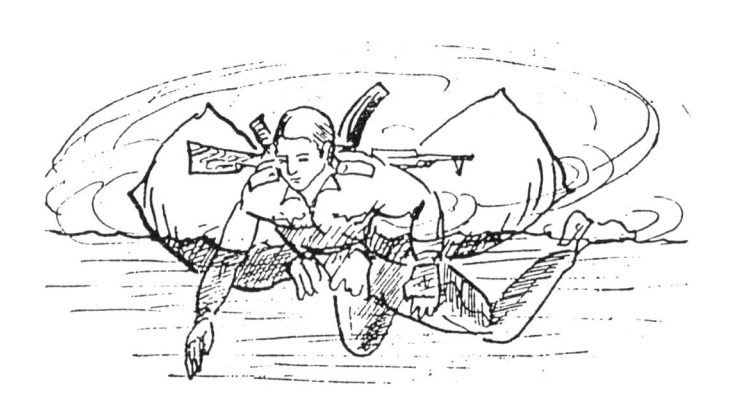

FIGURE 61. USING THE RAIN CAPE IN FORDING

MOVEMENT; OVERCOMING OBSTACLES; PENETRATING/STORMING BUILDINGS

If time and conditions allow, it is possible to use a rope, which is tied to pegs driven into the ground or tied around a tree (Figure 63). Little rafts can be strung out along the rope to help the weaker swimmers, the wounded, or prisoners whose limbs are bound. To transport heavy loads, individual rafts can be made from two or three plastic bags or other

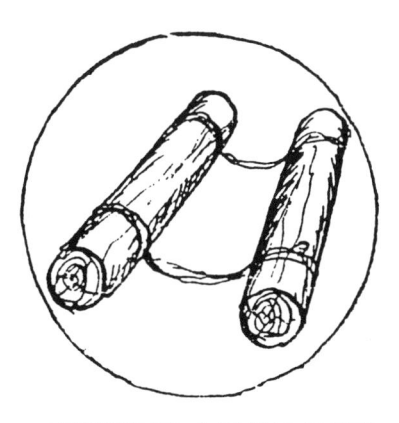

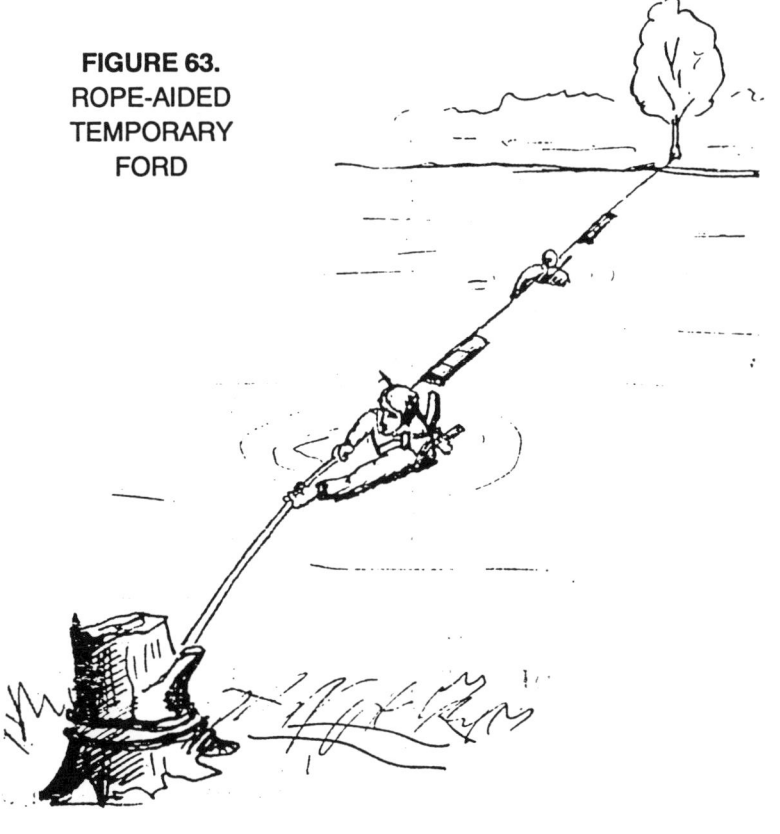

FIGURE 62. A SMALL RAFT FOR WEAK SWIMMERS

FIGURE 63. ROPE-AIDED TEMPORARY FORD

useful items, tying them together with rope or straps.

The raft form that provides the most carrying capacity is a three-cornered one that demands the least amount of lashing together (Figure 64).

During the teaching of fording to personnel, military discipline is to be maintained. It is absolutely necessary that a lifesaving team be assembled from among the strongest swimmers, as well as an inventory of lifesaving equipment: rings, ropes with floats, poles, and nets. There has to be an initial and concluding head count, by the roster, before and after any fording exercise.

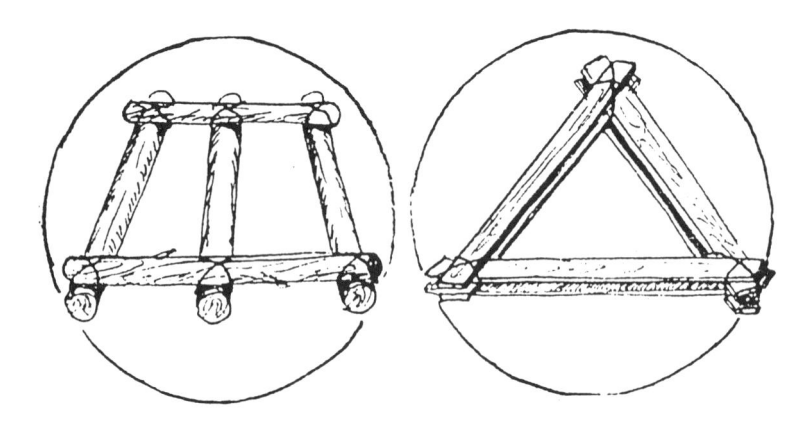

FIGURE 64. RAFTS FOR TRANSPORTING HEAVY LOADS

CHAPTER 3
TECHNIQUES AND METHODS FOR TEACHING PERSONAL COMBAT

The experience of the Great Fatherland War [World War II] showed that very few scouting patrols were good at personal combat (or capturing prisoners). Personal fighting by scouts is most often on a bold raid. It is the unexpected, silent attack on the enemy, the unforeseen encounter right in his own territory. In these operations grenades are thrown; weapons are fixed into positions; blows are delivered from cover with barrels, magazines, and butts; knives go into chests and backs; fists and feet are used—all to win the struggle and flatten the enemy.

This kind of combat, which may or may not use all kinds of personal weapons, develops a knowledge of defense and attack—of destroying a foe with a burst of automatic fire or a grenade; by bayonet, gun butt, knife, or entrenching tool; with the fingers, the fist, or knee; and by using incidental objects such as ropes, rocks, and so on. Mastering the use of "cold" [edged or pointed] weapons, useful objects, throws, choke holds (especially for capturing), or silent killing is done in actions at night, in woods, amid shattered buildings, and so on.

Depending on the mission, the personnel, equipment, features of an area, time of day, weather, and the enemy's movements and weaponry, an attack team or group uses any means in personal combat. A course in personal combat necessarily teaches the vulnerable points of the human body and some general and special exercises that can be used for conditioning the trainees physically. Lessons in special blows are an additional method of fostering single combat. These last are to be used as a complete set of exercises for morning exercise and in combination with running.

RECOMMENDATIONS FOR METHODS IN TEACHING TACTICS OF PERSONAL COMBAT

Practical instruction in personal combat tactics provides the ability to correctly assess conditions or individual situations and select the most effective action or technique or combination thereof. The whole process of teaching tactics may provisionally be divided into several stages, the basis for which is found in numerous repetitions of the tactics.

1. The first stage conforms to the aims of presenting a brief exposition of the characteristics and significance of the tactics, practicing the techniques in pairs, with one man defending passively, using various stances, hand/arm defenses, foul blows, and the attacker's inertia to beat him. It guides all defense or attack action to a logical end (e.g., stopping, pinning, disarming, tying up, or capturing). Through practicing with a partner, stances are learned, and both hands and sides are used, as is maneuvering around an opponent's static stance. Masking one's own movements and using psychology on an opponent and getting him (through feints) to make a mistake are also taught at this stage.

2. The second stage has the trainees learn the performance of defensive moves, to attack with blows and kicks in combination with throws once an opponent loses his balance, steadiness, or attention; how to make an opponent uncover himself by feinting at him; and using retreating, bending, twisting, and repelling in attack and defense, as well as employing physical objects.

3. The third stage teaches trainees how to use their weight, momentum, speed, and reflexes in attack and defense. It teaches how to complete a maneuver after being forced on the defensive

and demonstrates how the tempo of personal combat is affected by extra burdens, e.g., flak jackets, personal equipment, heavy boots, or snow on a street.

4. In the final stage, it is recommended that trainees learn single combat under increasingly complicated conditions (e.g., freeing oneself from an enemy's hold—in the midst of a chaotic situation). There should be instruction in carrying out attacks or defensive actions, using sound or visual signals. In the final analysis, it has to be emphasized that, in an unexpected encounter with the enemy, an attack has to be made with élan, using one or many of these methods. In mastering the art of personal combat, the teaching of the technical is closely tied to building up physical strength and agility and psychological conditioning.

BASIC VULNERABLE AREAS AND POINTS OF THE HUMAN BODY; TECHNIQUES FOR INFLICTING EFFECTIVE BLOWS

The points that are vulnerable to aimed blows are listed here by major parts of the body.

Head
• Frontal lobe area. Punches or blows with the edge of the hand produce shock and bleeding.
• Back of the head. A blow with a fist produces shock or death.
• Nose. A blow with a fist or edge of the hand produces fracturing, bleeding, or death.
• Bridge of the nose. A hammer blow with the fist produces fracturing and bleeding.
• Temple. A backhand blow, a punch, or a kick is fatal.
• Jaw area. A straight-on or side blow with a fist, knee, or foot brings on shock, concussion, fracturing, and pain.

Neck
• Carotid artery. A blow with the edge of the hand brings about shock or death.

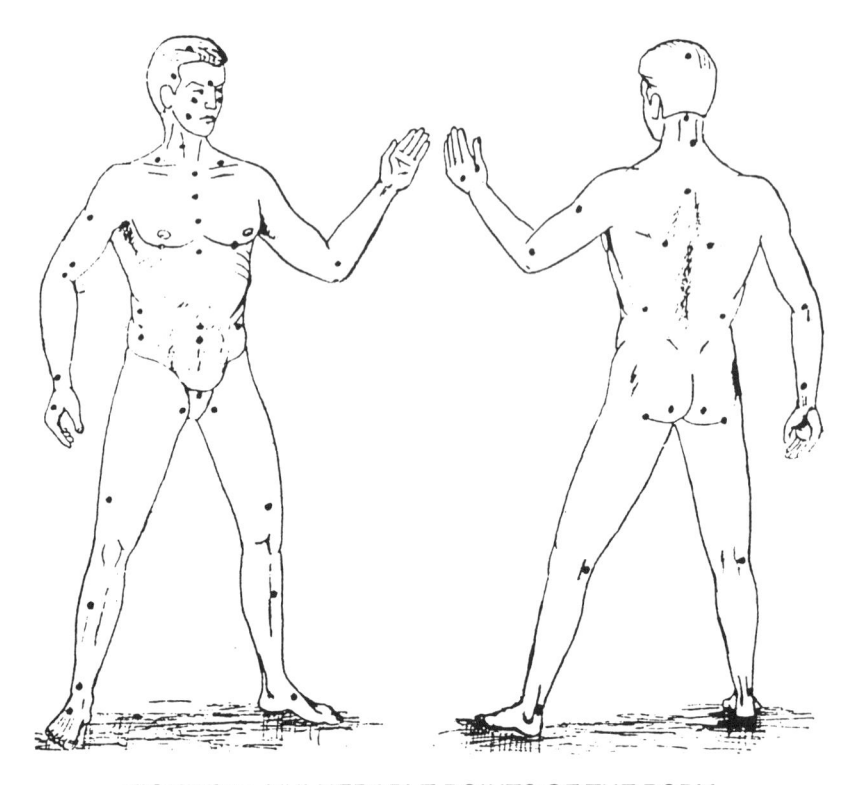

FIGURE 65. VULNERABLE POINTS OF THE BODY

• Esophagus. A blow with the edge of the hand or fist produces unbearable pain, bleeding, and/or death.

• Neck vertebrae. A blow from a fist, the two hands interlocked, or the edge of the hand produces fracturing, bleeding, or death.

Collarbone and Joints

• Collarbone. A blow with the side of the fist or edge of the hand brings about fracturing and sharp pain.

• Shoulder joint. A punch or an elbow produces extreme pain.

• Elbow joint. A punch or a kick produces extreme pain.

Stomach Region
• Solar plexus. A punch, jab with the fingers, or a kick with the knee or foot produces shock and constricting pain.
• Liver. A punch produces shock and sharp pain.
• Kidneys. A punch, a blow with the edge of the hand, a kick, or a stomp brings on sharp pain and urinary tract bleeding.
• Navel area. A jab with the fingers or elbow produces sharp pain.
• Groin. A blow with the side with the fist, a punch, an elbow, or an upward kick with foot or knee produces excruciating pain and shock.

Legs/Feet
• Knee joint. A kick with the toe or edge of the foot or the heel produces extreme pain.
• Shin. Kicking with the toe or scraping with the arch of the foot or a heel kick produces crippling pain, fracturing, or, sometimes, shock.
• Instep/arch. A stomp down on it produces sharp pain and fracturing.

The instructor has to be familiar with several exercises for general physical development that are performed in standing and sitting positions and by two-man teams. Naturally, warm-up exercises are of a rather general kind. Taking into account that the human body loses some of its flexibility after thirty years, some exercises presented here are for developing flexibility and increasing strength.

A GRADUATED SERIES OF WARM-UP EXERCISES

Warm-up exercises include those for the extremities: bending, squatting, push-ups, as well as those done by two-man teams (Figures 66 through 68). All moderate warm-up/stretching exercises are, as a rule, done after short runs. Repetitions vary in number from five to ten.

It is recommended that special exercises done in learning blows (which provides the means for attack and self-defense) be

KGB ALPHA TEAM TRAINING MANUAL

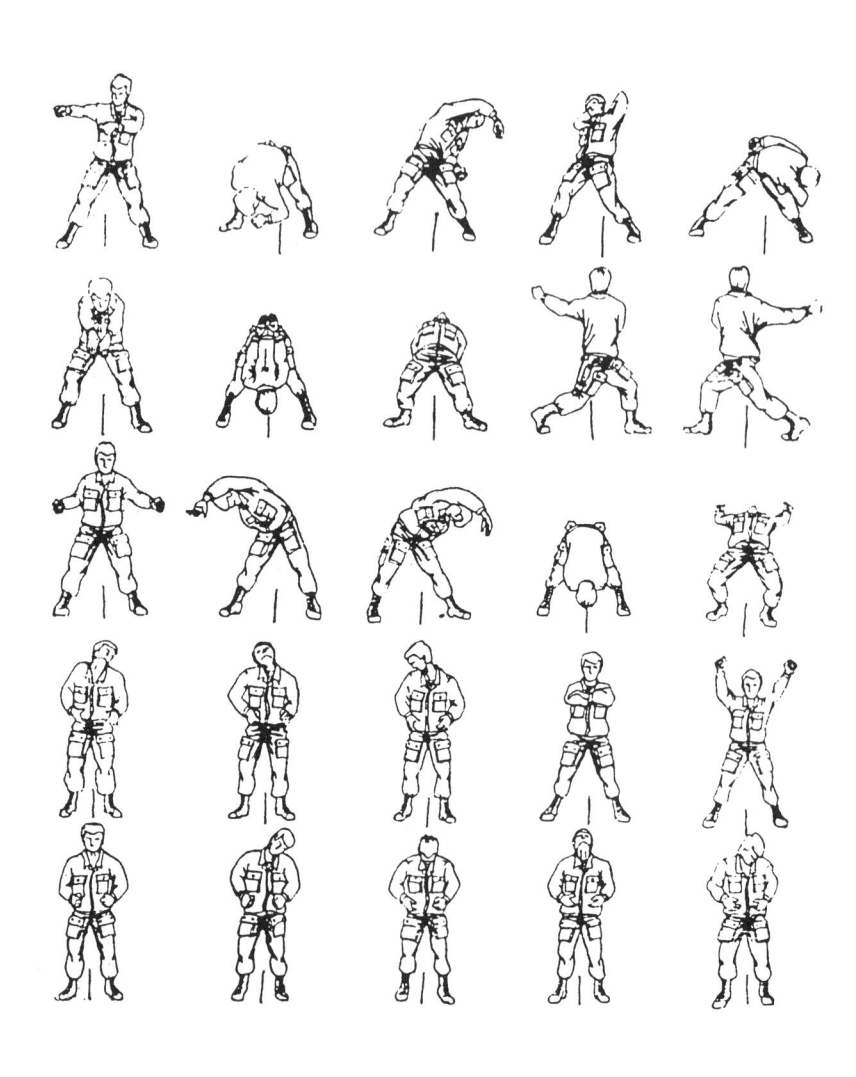

FIGURE 66. SERIES OF WARM-UP EXERCISES

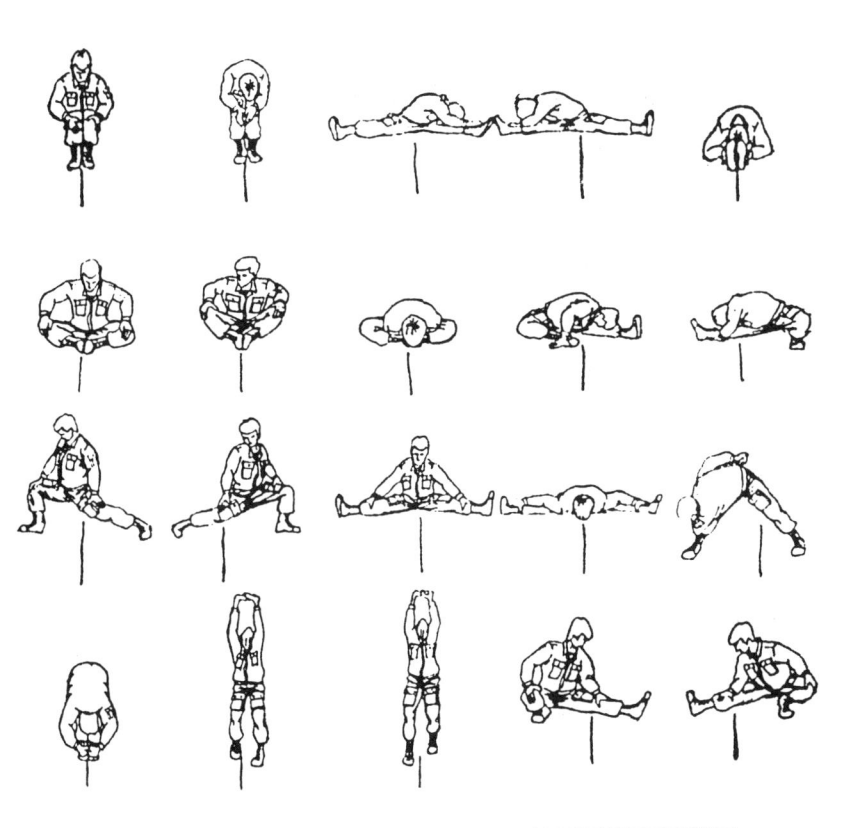

FIGURE 67. SERIES OF WARM-UP/STRETCHING EXERCISES FOR THE EXTREMITIES

performed daily by a complete unit or by individuals in their times away from regular duties.

SPECIAL EXERCISES

Defensive blocking using the arm with the forearm up, employed against hand blows, is shown in Figure 70. The initial position (IP)/basic stance (BS) is with the knees slightly bent, spread to about the shoulders' width; arms bent at the elbows, with the hands made into fists.

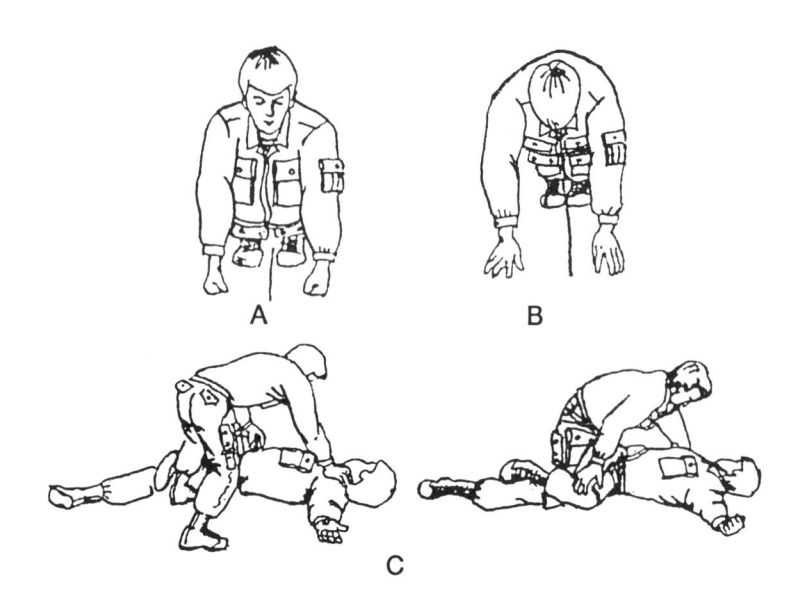

FIGURE 68. SERIES OF WARM-UP/STRETCHING EXERCISES
A, B—PUSH-UPS
C—EXERCISE BY A PAIR

Performance is as follows:

1. On "one," with a short, swift move, put the bent left arm slightly forward of the body with the forearm at about groin level.

2. On "two," switch the position of both arms so that the right one is positioned like the left in step 1—and the left one goes into the IP/BS. Repeat this ten times (Figure 71).

These are basic defense techniques used, most often, in all sorts of single combat against an armed or unarmed foe.

BLOWS

Blows with the Hands/Arms

A number of blows, which are performed in strict order, are described here.

TECHNIQUES AND METHODS FOR TEACHING PERSONAL COMBAT

FIGURE 69. EXERCISES DONE BY PAIRS

FIGURE 70. DEFENSE, WITH THE ARM AGAINST OVERHAND BLOWS

FIGURE 71. DEFENSE WITH THE FOREARM
AGAINST BLOWS FROM BELOW
A—AGAINST A FIST
B—AGAINST A KNEE
C—AGAINST A KNEE

• Elbow blows to the rear (Figure 72). IP is with the legs spread and the hands held pointing forward.

1. On "one," strike to the rear with one elbow, leaving the other bent and in place. Return the first arm to the IP. Repeat ten times.

• Elbow blows in sequence (Figure 73). IP/BS is with the legs slightly bent and the right elbow slightly forward.

FIGURE 72. ELBOWS BACK IP

1. On "one," hit forward with left elbow, bringing the right one back.

2. On "two," hit with the right elbow, returning the left to the IP. Repeat the sequence ten times.

• With the hands' edges (Figure 74). IP is with the feet at shoulders' width, hands behind the head.

1. On "one," strike the opponent's neck with both hands' edges simultaneously and then return to the IP. Repeat ten times.

• Punch (Figure 75). IP/BS is with the feet spread to

FIGURE 73. SEQUENTIAL ELBOW BLOW

FIGURE 74. SIMULTANEOUS BLOW WITH HANDS' EDGES

shoulders' width, fists held at the body's sides.

1. On "one," punch forward with the left hand.
2. On "two," punch forward with the right hand.

Repeat ten times, changing off on the hands used. This action is performed slowly, with tension, as though against resistance, and the breath is expelled with each blow.

Blows with the Feet/Legs

These blows are performed as described here.

FIGURE 75. PUNCH

• Swinging kick from below (Figure 76). IP/BS is with the feet slightly apart and the left arm extended in front with the palm down.

1. On "one," perform a straight-legged kick upward; strike with the left hand, and return to the IP. Repeat ten times.

Switch to the other leg and repeat the exercise ten times.

• Swinging kick back with the outside edge of the foot

FIGURE 76.
UPWARD KICK

FIGURE 77. KICK UP AND BACK

FIGURE 78.

(Figure 77). IP/BS is with the left foot pointed forward, hands made into fists and held low.

1. On "one," perform a straight-legged kick up and back, then bring the foot down to the IP. Repeat ten times.

Switch to the other leg and repeat the exercise ten times.

• Sideward kick with heel (Figure 78). IP/BS is with the arms down, slightly bent, the hands made into fists.

1. On "one," kick sideways and up with the left leg, and return to the IP. Repeat ten times.

Switch to the other leg and repeat the exercise ten times.

TECHNIQUES AND METHODS FOR TEACHING PERSONAL COMBAT

• Upward kick with the knee (Figure 79). IP/BS is with the left foot pointed forward, with the arms extended forward, and the breath held in.

1. On "one," kick the knee forward and upward, using the hands as if to pull the enemy's head down, at the same time breathing out; return to the IP. Repeat ten times.

Switch to the other leg and repeat the exercise ten times.

FIGURE 79. UPWARD KICK WITH THE KNEE

• Lashing kick "toward the chin" (Figure 80). IP/BS is with the hands down and forward into fists.

1. On "one," raise the knee and lash out with the lower leg to its full extension, and return to the IP. Repeat ten times.

Switch to the other leg and repeat the exercise ten times.

• Kick to the shin with the outside of the foot (Figure 81).

FIGURE 80. LASHING KICK

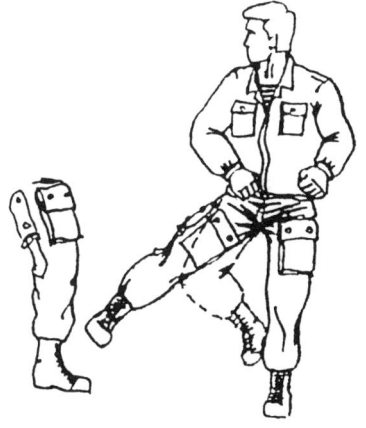

FIGURE 81. OUTWARD KICK TO THE SHIN

FIGURE 82. KICK WITH THE OUTSIDE OF THE ANKLE/SHIN

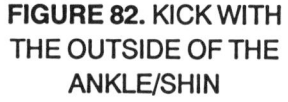

FIGURE 83. KICK WITH THE INSIDE OF THE ANKLE/SHIN

IP/BS is with the right foot positioned against the lower left leg, arms held low with the hands made into fists.

1. On "one," quickly kick the right leg to the side and down, and strike the air or ground. Repeat ten times.

Switch to the other leg and repeat the exercise ten times.

• Kick with the outside of the ankle/shin (Figure 82). IP/BS is with the hands held out to the side, palms back.

1. On "one," kick the right ankle/shin against the right palm and return to the IP. Repeat ten times.

Switch to the other leg and repeat the exercise ten times.

• Kick with the inside of the ankle/shin (Figure 83). IP/BS is with the feet spread to shoulders' width, the right hand held down with the palm down and parallel to the ground.

1. On "one," kick with the left inside of ankle/shin against the right palm and return to the IP. Repeat ten times.

Switch to the other leg and repeat ten times.

• Full swinging kick inward (Figure 84). IP/BS is with the left foot slightly to the rear and outside, with the right hand held forward, palm facing inward.

1. On "one," kick the left leg up and inward, hitting the right

TECHNIQUES AND METHODS FOR TEACHING PERSONAL COMBAT

palm with the inside of the ankle/shin, and return to the IP. Repeat ten times.

Switch to the other leg and repeat the exercise ten times.

• Full sideward kick with the outside of the foot/heel (Figure 85). IP/BS is with the arms slightly out to the sides with the hands made into fists.

1. On "one," raise the left knee and kick the leg out to the side at a 90-degree angle and return to the IP.

Repeat ten times.

Switch legs and repeat the exercise ten times.

• Forward kick to the chest or stomach with the bottom of the foot (Figure 86). IP is slightly to the rear with the foot resting on its toes.

1. On "one," kick the left foot out at the stomach; on "and," return it to the IP. Repeat ten times.

Switch to the other leg and repeat the exercise ten times.

• Backward kick with the heel (Figure 87). IP/BS is with the left leg slightly forward, the foot resting on the toe and the hands

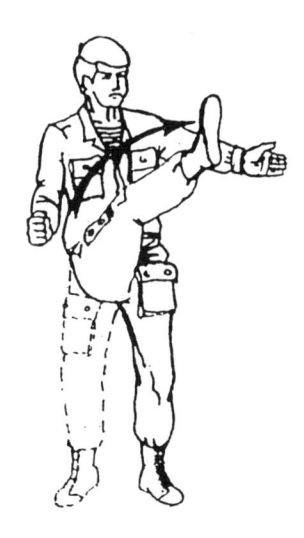

FIGURE 84. SWINGING KICK INWARD

FIGURE 85. SIDEWARD KICK WITH THE OUTSIDE OF THE FOOT/HEEL

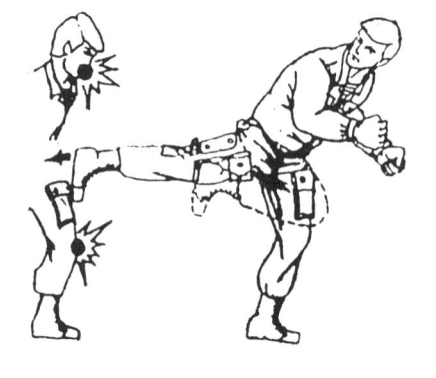

FIGURE 86. FORWARD KICK TO THE CHEST/STOMACH

FIGURE 87. BACKWARD KICK WITH THE HEEL

made into fists.

1. On "one," rest lightly on the right foot, lean forward a bit, and kick backward with the left, striking with the heel. Repeat ten times.

Switch to the other leg and repeat the exercise ten times.

• Forward kick with the outside of the foot (with a 90-degree turn) (Figure 88). IP/BS is with the left leg slightly forward and the hands made into fists.

1. On "one," kick forward with the outside of the left foot, at the same time turning the body 90 degrees to the right; return to the IP. Repeat ten times.

Switch to the other leg and repeat the exercise ten times.

• Swinging sideways, kick with the outside of the foot (Figure 89). IP/BS is with the left foot behind the right and the weight put on the right leg; the hands are held down, with the hands made into fists.

1. On "one," kick the left leg to the side and up, hitting with the bottom outside of the foot. Return to the IP. Repeat ten times.

Switch to the other leg and repeat the exercise ten times.

• Kick with the toe to the outside of the jaw (Figure 90). IP/BS is with the left foot to the rear, resting on its toe and the

FIGURE 88. FORWARD KICK WITH THE OUTSIDE OF THE FOOT

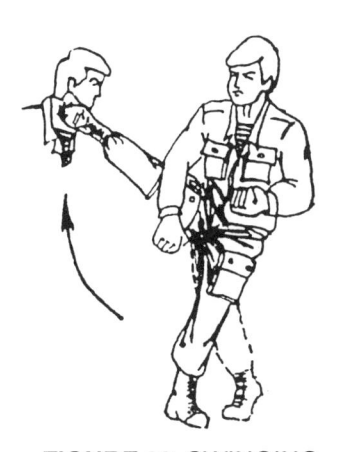

FIGURE 89. SWINGING SIDEWAYS KICK

FIGURE 90. KICK WITH THE TOE TO THE OUTSIDE OF THE JAW

hands made into fists.

1. On "one," raise the left knee and bend to the right, kicking forward and up with the foreleg to strike the jaw's side with the toe. Return to the IP. Repeat ten times.

Switch to the other leg and repeat the exercise ten times.

• Kicks delivered from a leap [flying kicks] (Figure 91). These are variants that are learned during the teaching of more complicated kicking methods (and practiced on a suspended punching bag). These are also good to learn to perfect physical coordination.

To strengthen the points of contact on the hands/arms and feet/legs (e.g., knuckles, hands' edges, elbows, heels, insteps), the following devices can be used "in combat": punching bags, stuffed

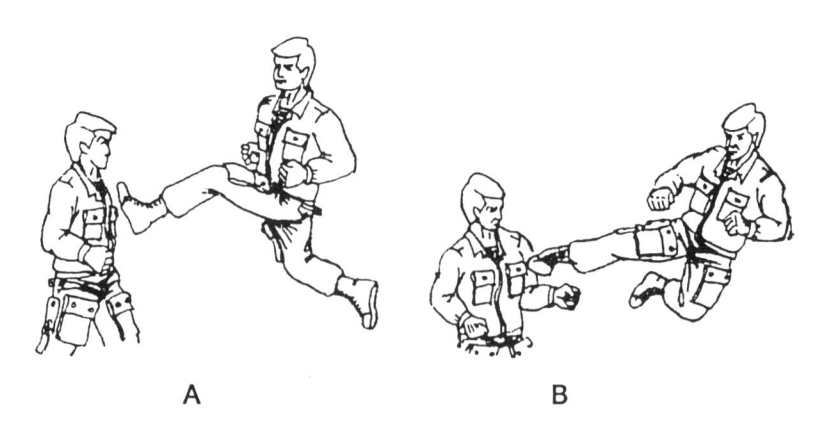

FIGURE 91. FLYING KICKS
A—STRAIGHT AHEAD
B—WITH THE SIDE

figures, dummies, and so on. These can be made from wood, bunched-up leather, rubber, straw, rope, and so on (Figure 92).

Special areas are to be used for developing the techniques of self-protection in falls onto one's side, back, and hands.

SAFETY AND SELF-PROTECTION IN FALLS

In every practice session it is essential to develop the techniques of correct ways of taking falls, with emphasis on safety and self-protection.

In performing any of these exercises, extreme care must be taken!

The following points must be observed:

• A partner's fall is lightened by the other partner supporting him when he is thrown.

• Throws are done only outward from the center toward the edge of the training area.

• Unlearned methods and/or throws are not performed.

• The thrown partner does not prop himself with his head or arm once he is down on the mat.

• Each instructor must diligently conduct warm-up exercises

TECHNIQUES AND METHODS FOR TEACHING PERSONAL COMBAT

FIGURE 92. "COMBAT" EXERCISES
WITH HANDS, ARMS, FEET

for the trainees, and know and follow safety procedures.

Safety/self-protection in taking falls onto the back begins with knowing the final position that the one falling will end up in. This means practice. He lies on his back, draws his knees back to his chest, and extends his arms outward on the floor (Figure 93).

Without straightening his legs, he rocks forward a few times, slapping his extended arms and hands on the floor. After this, he learns how to fall to this final position from an upright sitting position. In this sitting position, he extends his arms straight forward, with his chin on his chest, and falls onto his back, slapping the floor. In this fall, he goes through a swimming-like motion, drawing the knees to the chest. The back of his head does not touch the floor, the palms do the slapping, and the arms are held at 45 degrees out from the body (Figure 93).

To come quickly to a combat-ready stance without falling

FIGURE 93. FALLING ON THE BACK

FIGURE 94. THRUST WITH THE LEGS

back again, use your personal weapon, and dig your fingers into the sand, grass, or dirt that you may be lying on. At the same time, pull the knees back to the chest and then thrust the legs straight forward to jump to an upright position. A starting position is show in Figure 94. (A hastily scooped up handful of sand or gravel thrown in the face can help repel an enemy in this situation.)

During practice sessions, safety in falls can be increased with the partner's taking hold of the faller's legs, as shown in Figure 95.

FIGURE 95. A PARTNER HELPING IN A FALL ON THE BACK

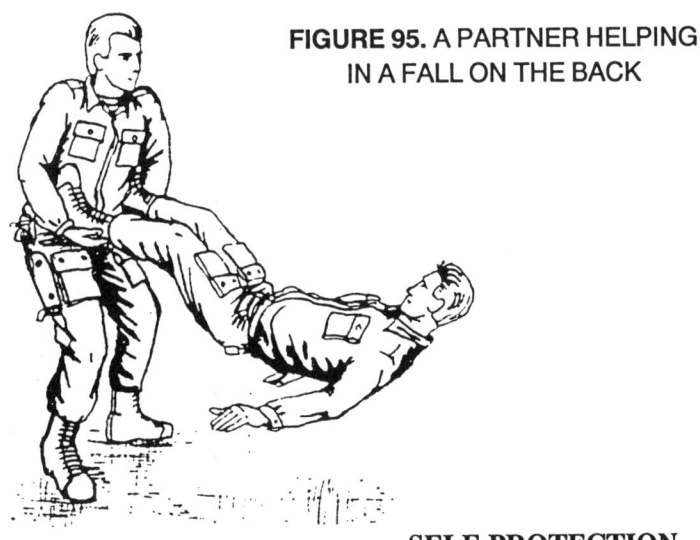

FIGURE 96.

SELF-PROTECTION IN TAKING FALLS TO THE SIDE

This type of self-protection begins with mastering the final position that you must assume after a fall. You have to learn how to come to this position (Figure 96) from a supine position. You have to slap the floor with one extended hand and pull the knees up to the chest. After this, you put the bent leg on the same side as the slapping hand, in line with that hand about the mid-point of the other leg, which now lies straight on the floor.

To make getting into this position more effective, roll from side to side on your back.

FIGURE 97. SELF-PROTECTION IN TRIPPING, THROWING

Once this position is learned, you should learn how to fall safely from the sitting position and into the leaning rest position. During any of these, the head should not be thrown back.

To improve falling correctly onto your side, being tripped and thrown backward by an opponent can be used (Figure 97).

SAFETY IN FORWARD FALLS

The set of exercises for quickly learning this kind of self-protection is shown in Figure 98. The IP is the leaning rest position (with palms flat).
• Flex and push with the fingers
• Flex and unflex and slap the palms down
A further set of exercises includes the following:
• Fall forward onto the hands from an IP of kneeling upright.
• Fall forward onto the hands from an IP of standing upright.
In falling forward, cup your hands very slightly to absorb the impact. Do not fall onto flattened hands.

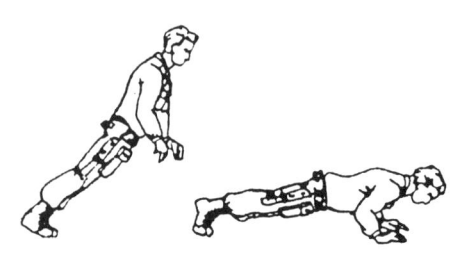

FIGURE 98. SELF-PROTECTION IN FALLING FORWARD

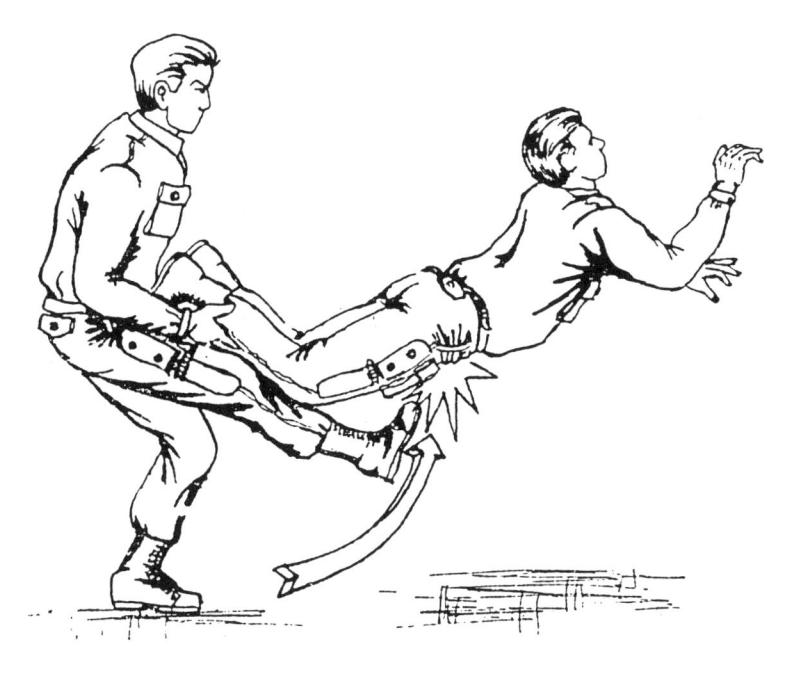

FIGURE 99. SAFETY IN FALLS BY HAVING THE LEGS HELD

To ensure a correct fall when a team of two is working on these exercises, turns can be taken in falling and holding each other's legs. Holding the legs/ankles lessens the fall's impact (Figure 99).

A SERIES OF EXERCISES IN LEARNING SAFETY/SELF-PROTECTION

Forward and Backward Falls

The final position is supine with the arms out from the body at an acute angle, palms flat on the floor, legs bent and crossed at the ankles.

• Rock back and forth in the above final position on your back.

• Roll backward from a squat with the buttocks touching the floor.

• Fall backward, using a partner's back.

- Fall onto the back from a standing position.

Throws are never attempted in the first lessons!

Sideward Falls

The final position is lying on the side.
- Roll from left to right in the final position.
- Fall from a squat onto the buttocks and side.
- Fall from a hop without the buttocks touching the floor.
- Fall with the help of a partner or a pole.
- Fall from a standing position.

To improve performance and safety, use these falls and a backward sloping ramp.

Forward Falls

The final position is the leaning rest position.
- Flex the hands while in the leaning rest position.
- Fall onto slightly extended, slightly cupped hands from a kneeling position.
- Roll onto the chest and stomach from the kneeling position, with arms behind the back and head turned to one side.
- Fall in place.
- Fall with thrusts ("jerks") forward.

Have a partner hold the feet to improve safety in performance.

CHAPTER 4
A Practical Section in Special Physical Training

The attack is the basis of action in capturing or killing an enemy with the help of various physical objects and "cold" weapons. This body of instruction is in two parts: methods for capturing and methods for killing.

BASIC METHODS FOR CAPTURING

The goal of the activities and techniques presented in this section is the capture and consequent transport of enemy soldiers. There is a considerable amount of literature on methods of capture, but, in the end, there remains one fundamental fact: taking an armed enemy is done only by a team—of no less than two persons. One person alone cannot accomplish this; there has to be close armed support from another one. This support comprises using blows for defense or silencing, using one's hands or feet, or ultimately using a weapon in a critical moment. This manual examines the most effective methods, which have been refined from their application in real circumstances, against real armed enemies. By and large, these methods are connected with the tactics of concealed movement and actions in taking objectives from the rear.

The capturing methods include, e.g., using throws, kicks, or prisoner-securing techniques, which have to be performed in an exceptionally short period of time, and as a one-time action. If an attempt to get close to the enemy doesn't work, then, necessarily, one has to wait for the enemy to come. The enemy's weapons have to be neutralized, obviating their use, by using, for instance, blows to his shoulder with a weapon or to the back of his head with a fist or a rock.

Almost anything—sleeves, shirts, rags, sacks, rain capes—can be thrown over his head to muffle his cries. The attacking team does not just plan who will do what in the upcoming action: grabbing and tying up a prisoner, suppressing resistance, escorting the captives away. The team has to practice; rehearse to the point that all actions are done automatically. These are fundamental requisites for silently capturing an enemy.

Arm Hold in Downing and Seizing an Enemy from Behind

A grab and a sharp blow from behind to an enemy's arm are used to bring him down, using his own weight. The steps are shown in Figures 100 and 101.

The real dynamics are in the approach from behind and

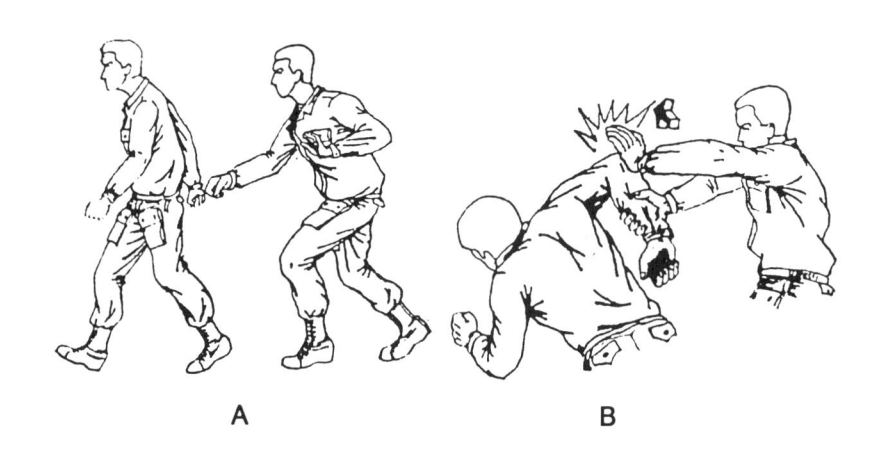

FIGURE 100. ARM HOLD IN AN ATTACK FROM BEHIND
A— APPROACH
B—HOLD AND BLOW

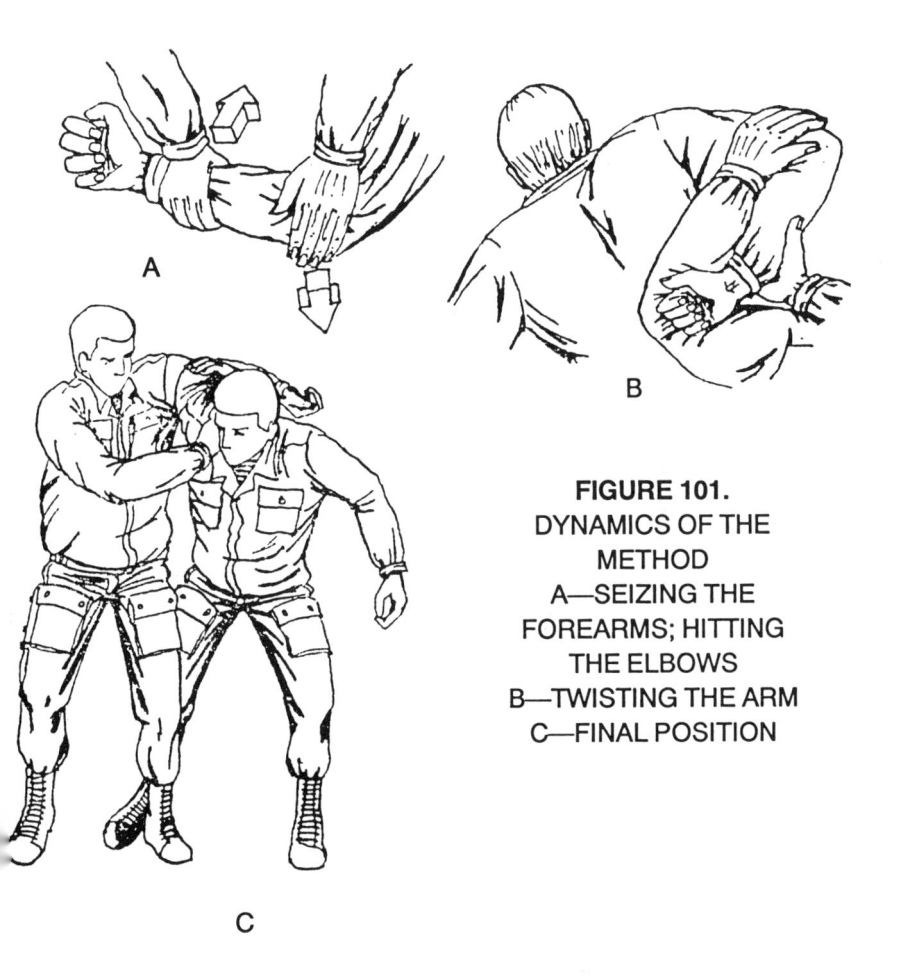

FIGURE 101.
DYNAMICS OF THE METHOD
A—SEIZING THE FOREARMS; HITTING THE ELBOWS
B—TWISTING THE ARM
C—FINAL POSITION

twisting the arm. Fundamental to this method is the strong blow to the elbow. Additional dynamics are shown in Figure 101. After all this, a sharp thrust of the hand to the shoulder takes him down for securing and conveying away.

The most efficient and successful techniques for silent capturing are surveyed in the following sections.

Under combat conditions, approaching an enemy from behind and stunning him with a blow to the head with an assault rifle can be used, followed by twisting his left arm behind his back and taking him to the ground.

Chokehold with the Arm

This method goes as follows: the first member of a two-man team grabs an enemy from behind (Figure 102) in a choke hold, and taking his right hand in his left, tightens the hold. The second team member immediately grabs the enemy's legs around the calves, and both carry the captive off to cover or to a vehicle.

A sack, tunic, or jacket can first be thrown over the enemy's

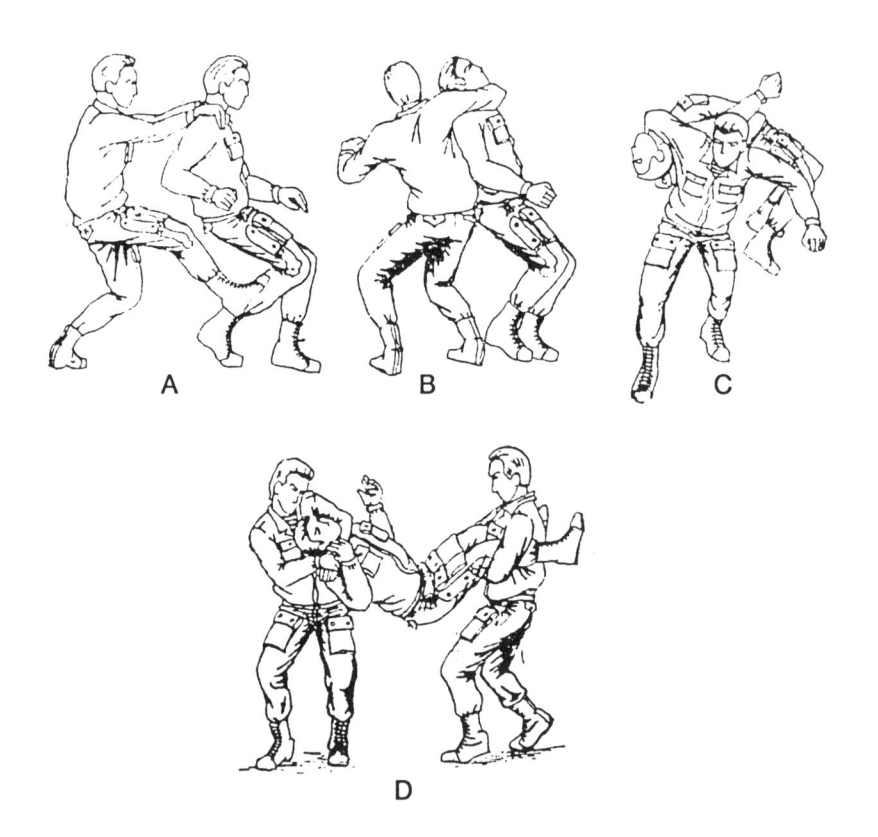

FIGURE 102. CHOKE HOLD WITH THE ARM
A—GRABBING THE SHOULDERS A KICK BEHIND THE KNEE
B—HOLDING THE NECK AND THRWING HIM TO THE SIDE
C—TRANSFER TO THE BACK
D—TRANSPORT BY A TEAM OF TWO

head and then the chokehold applied. Strong pressure should be used to cut off any shout but not to suffocate the enemy before moving him out. Watch out for a weapon in his hand.

Taking an Enemy by Grabbing His Legs from Behind

The technique is to come up behind the enemy and jerk his legs out from under him, simultaneously kicking him as he falls forward (Figure 103).

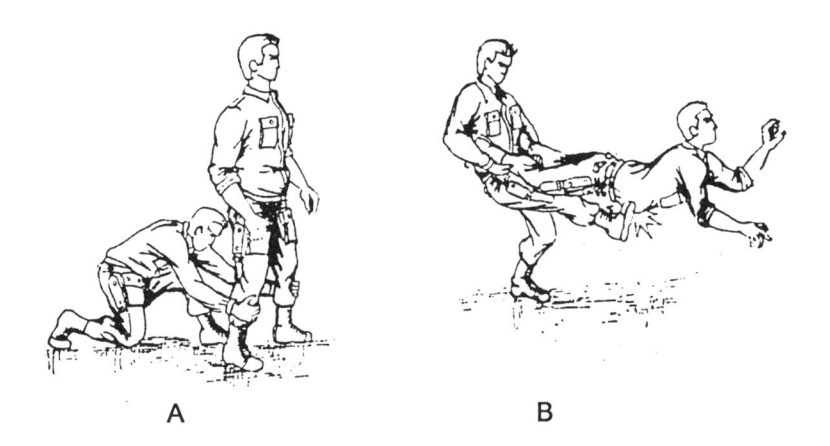

FIGURE 103. SEIZING THE LEGS FROM BEHIND
A—SEIZING
B—THROWING

After he is down, raise him up to gag him, stun him, or tie his hands behind him. Then take him off to wherever is decided. Conveying a prisoner is done by staying at his side ("hand in hand"). If a silent attempt goes wrong and you are spotted, you have to deliver a blow to the enemy's face (eyes or nose) and carry out the kick and throw shown in Figure 104. Then you have to turn him face down, get him secured and moved out as already determined. This method works well in woods and out-of-the way spots, or even on the street when taking particularly dangerous enemies.

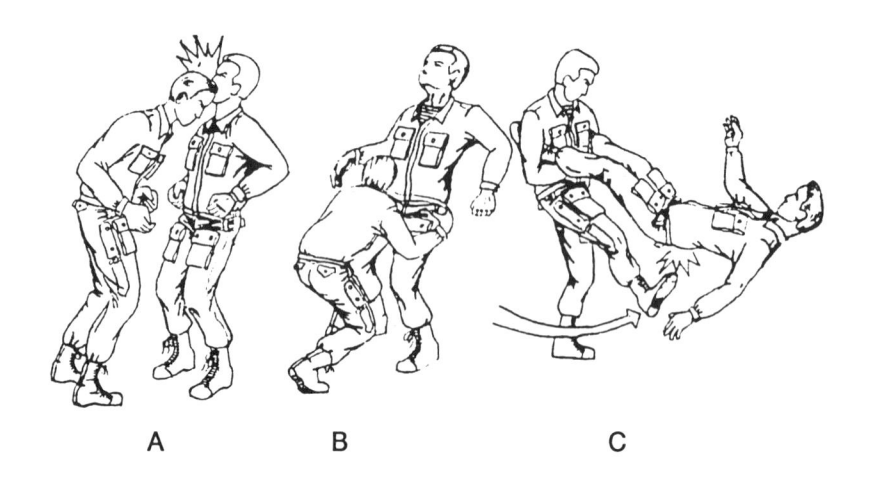

FIGURE 104. THROW BY SEIZING THE LEGS FROM IN FRONT
A—A BUTT IN THE FACE
B—GRABBING THE LEGS
C—THE THROW

Attack from Behind, Grabbing the Crotch

This method is hardly humane, but it is extraordinarily simple and reliable. It is used under any conditions, but especially in a tight situation in an urban area against an armed and dangerous foe, regardless of his strength, size, or weapon.

The way to do it is to get right up to him by coming out of concealment (from a crowd or angle of a building), attack, and get him to the designated destination. The technique is shown in Figure 105. Use the right hand to grab the crotch, the left to grab the shoulder or collar. Pull backward with the right hand while pushing forward with the left.

If he tears loose, trip him or knock him down (Figure 106).

Capturing by Using a Sack, Rain Cape, or Tunic over the Head

A chokehold is put on the neck and throat with the arms. The attack is from behind. It is appropriate to use capturing methods already described for action in wooded or hilly locales—e.g., on

A PRACTICAL SECTION IN SPECIAL PHYSICAL TRAINING

A

B

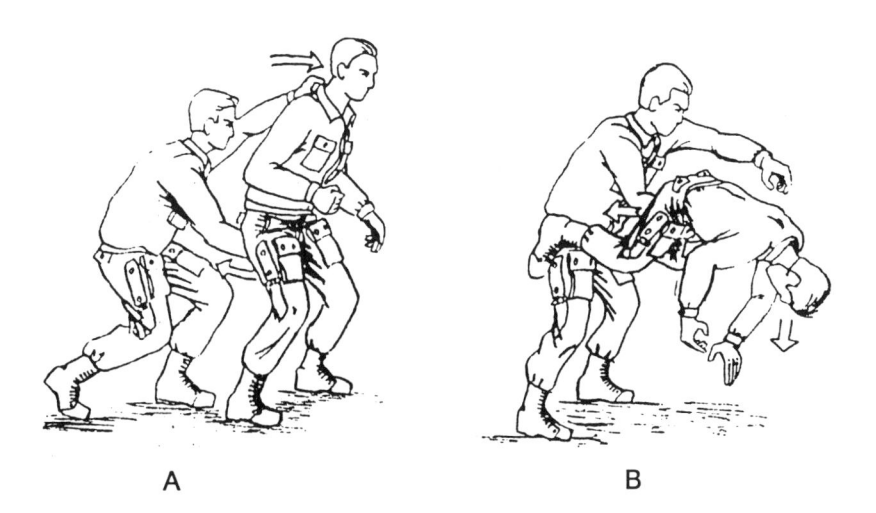

FIGURE 105. GRABBING THE CROTCH FROM BEHIND
A—GRAB
B—THROW

FIGURE 106. TRIPPING A FLEEING ENEMY

FIGURE 107. CAPTURE USING A RAIN CAPE
A—ALONE
B—IN A PAIR

tracks, in gorges, by streams and springs (Figure 107).

Additional Methods
• The "Hand-in-Hand" Technique.

Variant 1

The hand-in-hand hold is done on the enemy's wrist (Figure 108). On a street, one or two individuals come up to the target from behind, grabbing his forearm and bending down his hand at the wrist or fingers, forcing his fingers back toward the wrist.

Continuing with his right hand, a capturer takes the target's hand, keeping its back upward, and bends it downward and under at the wrist. Thus inflicting sharp pain, the capturer(s) take(s) the prisoner away.

Variant 2

As shown in Figure 109, the capturing action is analogous to that of Variant 1. This method, though, inflicts pain by twisting

A PRACTICAL SECTION IN SPECIAL PHYSICAL TRAINING

A B

FIGURE 108. "HAND-IN-HAND" CAPTURE: THE WRIST
A—START
B—FINISH

A B

FIGURE 109. "HAND-IN-HAND" CAPTURE: THE FINGERS
A—START
B—FINISH

the hand over and bending the fingers down and back toward the target's body.

When either variant is used, the second member of the team should stay on the prisoner's other side, a bit to the rear, in case the target has a weapon in his pocket.

In an urban setting, a variation of this method has the capturer keeping his lock on the prisoner's hand and also holding him by the back of his belt—in case of hidden weapons (Figure 110). In this, the other (armed) partner, must keep behind or to the side of the prisoner. This method is used in crowds, which allows quickly seizing and pushing the target into a vehicle. The other partner stays right behind or near the spot of capture. Both partners must act smoothly and in concert.

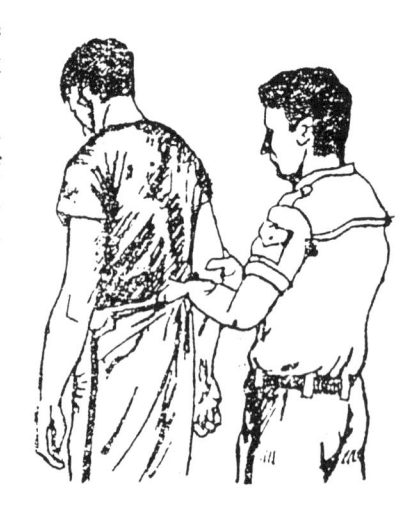

FIGURE 110. SEIZING THE BELT (WAISTBAND)

Capturing a Seated Enemy

Go right up behind the target seated on a chair and put a full-arm chokehold on him.

The other partner grabs the target's feet and helps carry him out of the room, and keeps a weapon ready in his free hand (Figure 111).

A prisoner taken inside a room has to be pushed out the door very quickly. A special technique is needed for this. During the capturing action, using a choke hold, transfer the hold down to the prisoner's groin (Figure 112). This technique is effective even though it is exceptionally brutal.

In urban operations, you have to know how to capture an enemy sitting in a car. Two methods are given here.

A PRACTICAL SECTION IN SPECIAL PHYSICAL TRAINING

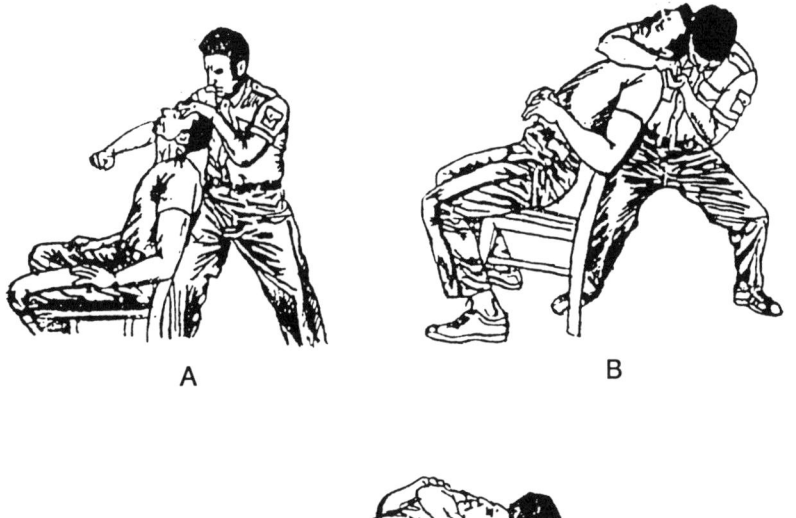

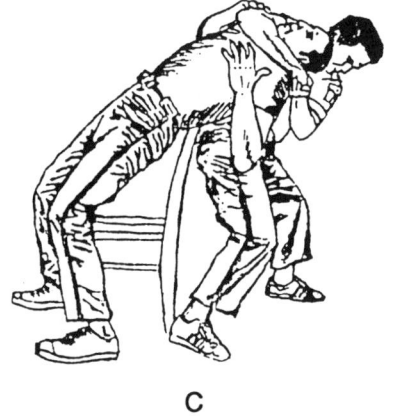

FIGURE 111. CAPTURE OFF A CHAIR IN A ROOM
A—PULLING BACK THE HEAD
B—APPLYING THE CHOKEHOLD
C—LIFTING THE TARGET FROMT HE CHAIR

First Method

The basic attack is aimed at the throat, neck, and head. After opening the driver's-side door, put your arm around the driver's

KGB ALPHA TEAM TRAINING MANUAL

A B

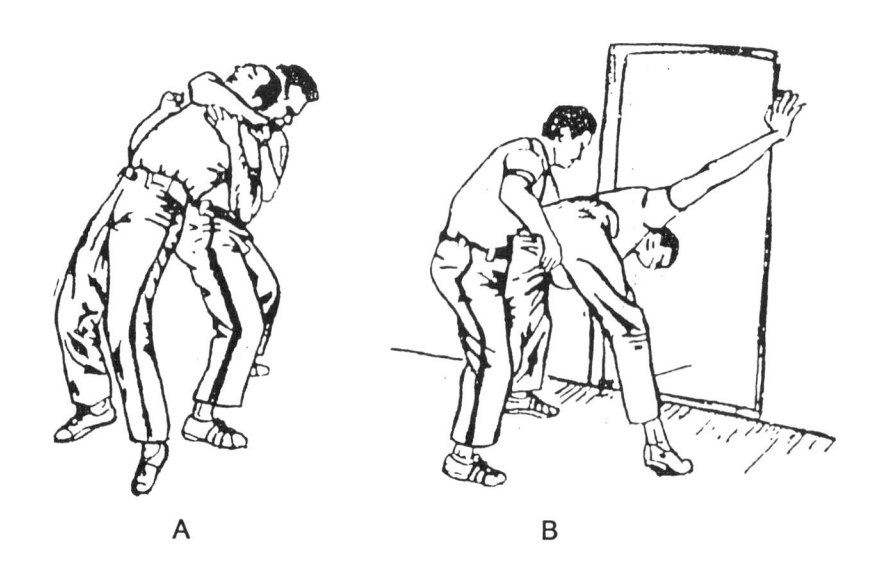

FIGURE 112. TAKING A PRISONER OUT A DOOR
A—STRUGGLE
B—CROTCH TWIST

A B

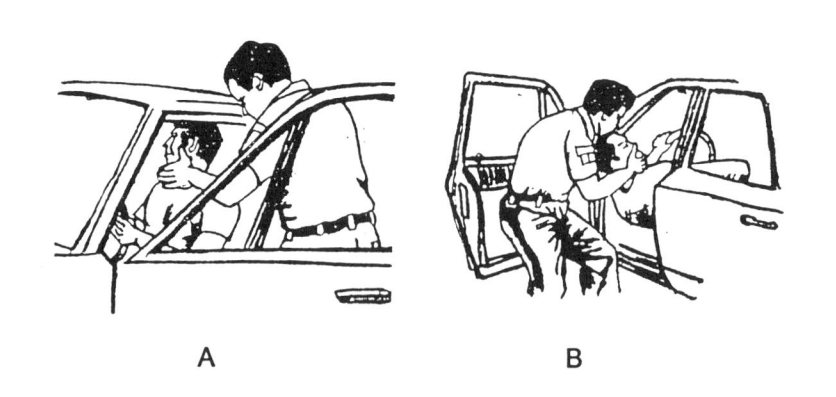

FIGURE 113. CAPTURING AN ENEMY IN A CAR
A—GRABBING THE HEAD
B—CHOKEHOLD AND EXTRACTION

A PRACTICAL SECTION IN SPECIAL PHYSICAL TRAINING

throat and pull him out and take him away. The same can be done by bending the driver's head back and pulling him out by holding his chin (Figure 113).

Second Method

This is based on hurting the enemy's arm. Start with a blow to his arm with the edge of your hand, and then start to bend his hand in the hand-in-hand manner or twist his arm behind his back. To complete this action, pull him out of the car, applying

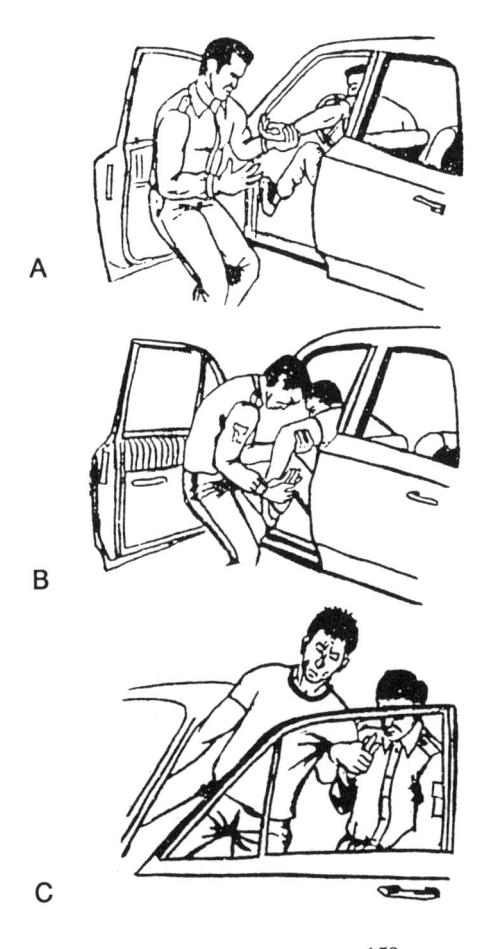

FIGURE 114. CAPTURING AN ENEMY IN A CAR
A—TWISTING THE ARM
B—BENDING THE HAND
C—EXTRACTION, APPLYING REQUISITE PAIN

KGB ALPHA TEAM TRAINING MANUAL

**FIGURE 115.
CAPTURING AN
ENEMY IN A CAR
A—SEIZING
THE HAND
B—TWISTING
THE HAND
C—TWISTING
THE HAND**

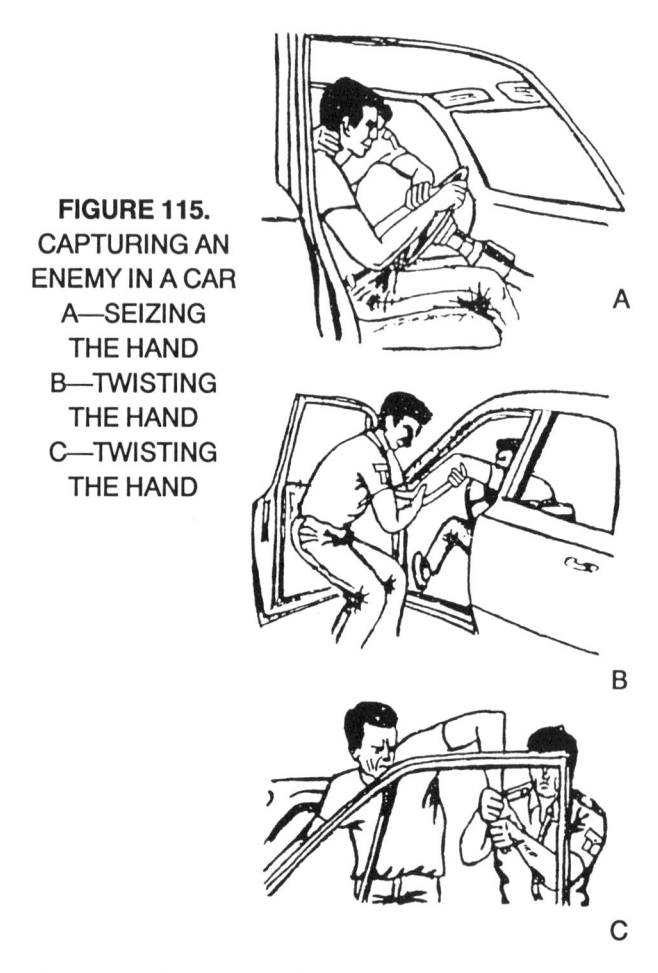

whatever pain is needed to make him cooperate (Figures 114 and 115).

BASIC METHODS FOR SILENTLY KILLING AN ARMED ENEMY

Silent elimination (killing) of an armed enemy may be done by noncontact methods, that is, with silent weapons. However, using such weapons, especially in the dark, does not guarantee a sure kill. There may be a miss or only wounding, which leads to noise and a resulting discovery of the attempt.

The surest way to silently eliminate an armed enemy is with direct physical contact.

Method 1

After stealing up behind the target, deliver a powerful blow to his head with a rifle butt (or hammer, ax, or heavy object). Immediately after, put your hand over his mouth and bring him down to the ground. Your team then acts according to plan.

Method 2

Steal up behind the enemy or stalk him down a path. Put your left hand over his mouth and cut his throat with the knife in your hand (Figure 116). Let this dead sentry fall to the ground or drag him off to the side.

Another way of taking out an enemy is to jump him at a run from behind. To do this silently and stealthily is not possible because of all the activity involved. After jumping him, use a chokehold or cover his mouth with your hand. A kick to the crotch works, too. The enemy will fall forward or to his side from a blow from the rear. Use a knife on his throat, as shown in Figure 116, or stab him up under the rib cage or between his ribs in the heart region. But cutting the throat is the surest method: a flak jacket provides no defense against this. The technique is shown in Figure 117.

Stabbing the throat is learned for security's sake, in the event that a targeted enemy turns toward the direction of the attack. Right after a stab, you have to grapple with the enemy and get him down on the ground, shut his mouth, and—most important—keep him from firing his weapon.

If the footing is unsure or the space constricted, a chokehold can be used, with the attacker falling to the ground with the target but staying on the target's back. Physical objects can be used to aid in choking: e.g., a short length of rope (garrote), wire, or a stick. If it looks like choking will not work, use a direct blow to his head with a hard object (e.g., a rock, a stone, or a chunk of asphalt).

KGB ALPHA TEAM TRAINING MANUAL

FIGURE 116. SILENT DESTRUCTION OF AN ENEMY
A—APPROACHING
B—COVERING THE MOUTH
C—BENDING HIM BACK
D—KILLING

A PRACTICAL SECTION IN SPECIAL PHYSICAL TRAINING

FIGURE 117. KILLING BY JUMPING FROM BEHIND

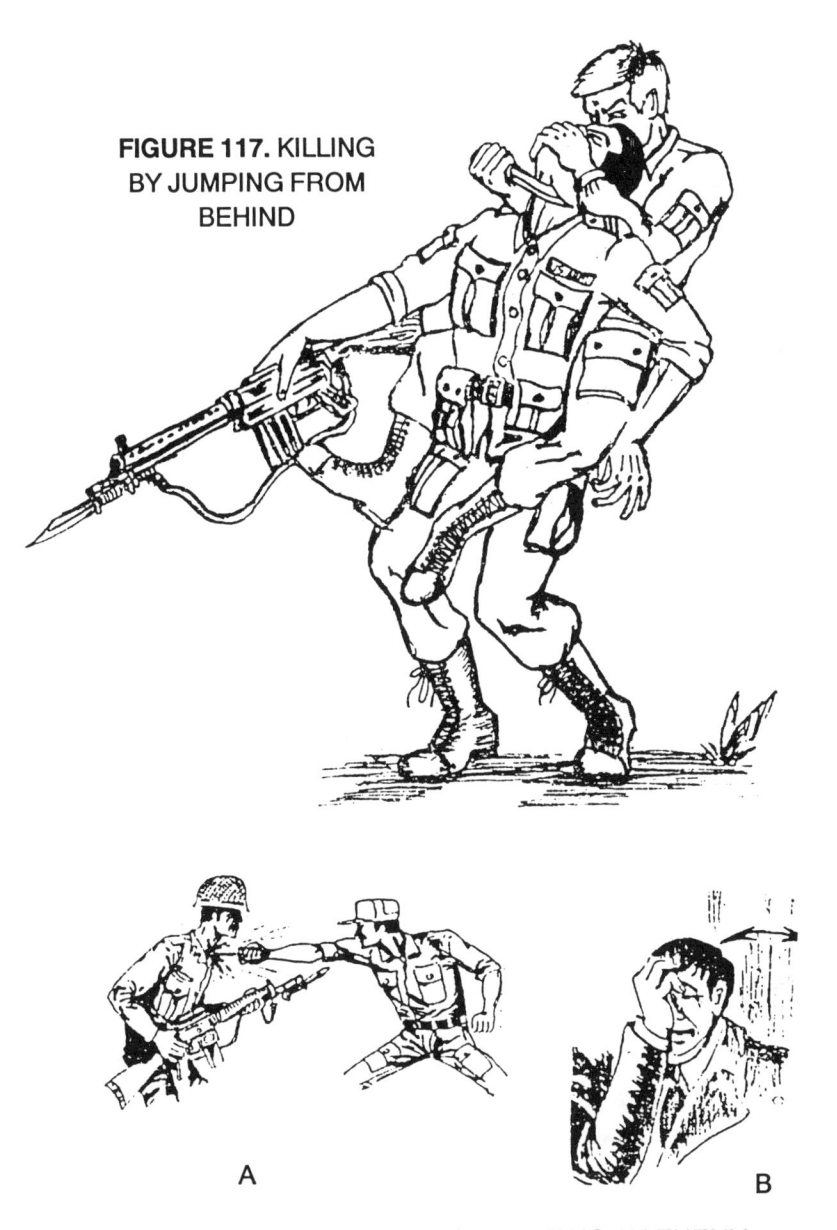

FIGURE 118. SILENTLY ELIMINATING AN ENEMY
A—STAB TO THE THROAT
B—BLOW TO THE HEAD WITH A HARD OBJECT

FIGURE 119. CHOKING METHOD
A—APPROACHING
B—JUMPING
C—SEIZING
D—CHOKING

The other partner can hit the enemy's head (on the forehead or crown) with a weapon immediately after the first partner has attacked, or he can stab the enemy in, say, the heart.

Method 3

Come up from behind and throw a garrote around the enemy's neck, swiftly pull him to your back, and pull the rope's

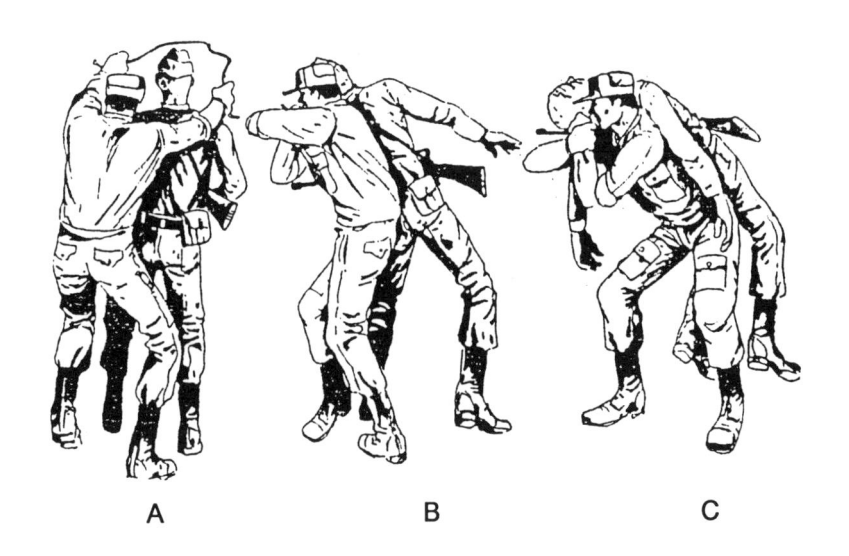

FIGURE 120. CHOKING WITH A ROPE (GARROTE)
A—THROWING ON THE ROPE
B—TURNING TO THE LEFT
C—CARRYING THE ENEMY ON THE BACK

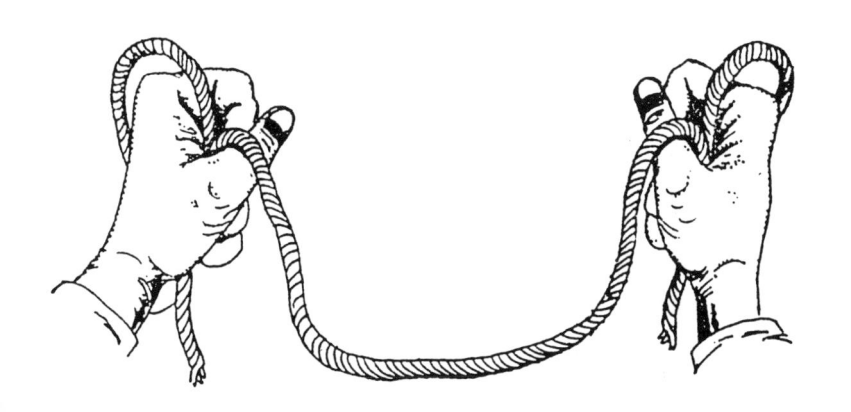

FIGURE 121. CHOKING ROPE (GARROTE)

ends over your shoulder (Figure 120). The rope aids in dragging the enemy sentry into cover (e.g., around a corner, behind bushes, into a vehicle).

This method is very reliable. Evading this attack is impossible, as is shouting. The implement used is shown in Figure 121. The length of rope between the hands is thirty to fifty centimeters.

As a rule, when it is impossible to come up behind and unseen to an enemy/sentry, a cold weapon (e.g., ax, sharp knife) can be thrown at him from concealment at his rear. In this way, the intervening "dead" ground can be covered and the sentry finished with, say, a bayonet (or rifle butt or knife), as in Figure 122.

These are basic methods and techniques for killing silently. Techniques for throwing cold weapons are described in subsequent sections.

There are many reliable ways of killing an enemy silently. In various situations, they can be used for self-protection, and thus they conform to the teaching of that to trainees.

ADDITIONAL METHODS FOR SILENT KILLING

Cold Weapons

Figures 123 through 125 show how to deal mortal stabs in various locales under combat conditions.

When a knife is not used in a kill, different choking methods can be used. They can be done in direct-contact, one-to-one combat: e.g., in unexpected encoun-

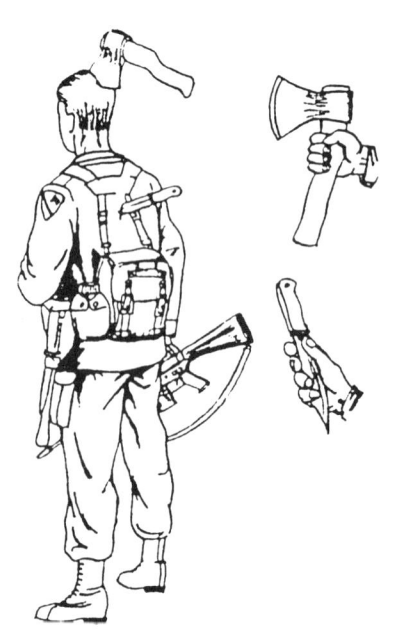

FIGURE 122. KILLING AN ENEMY FROM BEHIND WITH A COLD WEAPON

A B

FIGURE 123. STABBING WHILE
COVERING AN ENEMY'S MOUTH
A—UNDERHAND STAB
B—OVERHAND STAB

A B

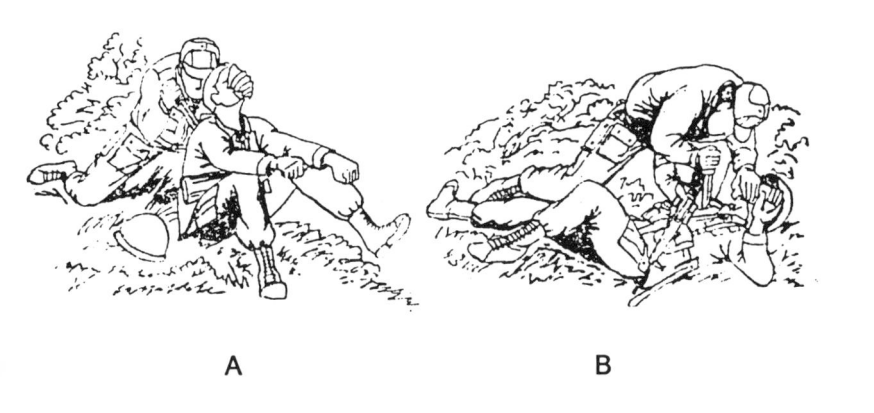

FIGURE 124. ATTACKING AND STABBING AN ENEMY

FIGURE 125. USING A KNIFE IN CLOSE-UP FIGHTING
A—BACKHAND SLASH TO THE FACE
B—OVERHAND STAB TO THE CHEST
C—UNDERHAND STAB

ters, in capturing a standing or supine foe, or in evading capture.

CHOKING TECHNIQUES

Choking by Grabbing the Collar

From the front, grab the enemy's collar on each side with a cross-hand hold. Twist your clenched hands inward, using the thumb knuckles to press hard against his carotid arteries or esophagus (Figure 126).

A variation on this is shown in Figure 127.

Choking in Fighting Down on the Ground

Knock your foe down, hold on to him, and choke him.

FIGURE 126. CHOKING BY GRABBING THE COLLAR'S SIDES

A PRACTICAL SECTION IN SPECIAL PHYSICAL TRAINING

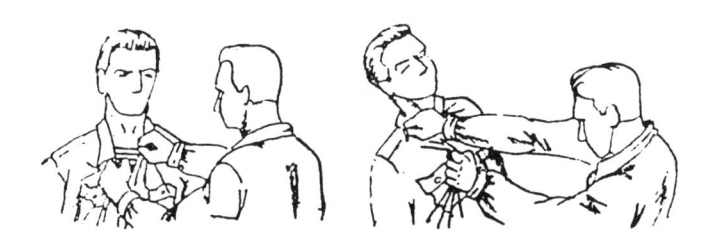

FIGURE 127. CHOKING WITH THE COLLAR

Choking with Your Forearm While Pinning an Enemy Down with Your Side

Seize the enemy's collar and press down heavily on his throat with your forearm, simultaneously pressing him to the ground with your side and not letting him get loose (Figure 128).

FIGURE 128. CHOKING WITH THE FOREARM

Choking with the Legs

As soon as you have gotten your foe down, quickly grab him by the hair and, yanking it upward, straddle his neck (Figure 129). Fall to the side without letting go of his head, and squeeze his neck with your legs straightened and crossed at the ankle until he loses consciousness.

Hooking the nostrils or upper lip can be done instead of grabbing the hair. Pull the head high enough so that the neck can be caught between your legs.

Choking with the Fingers

This method can be used either when standing or lying on the ground. From either the front or the side, grab the throat in the

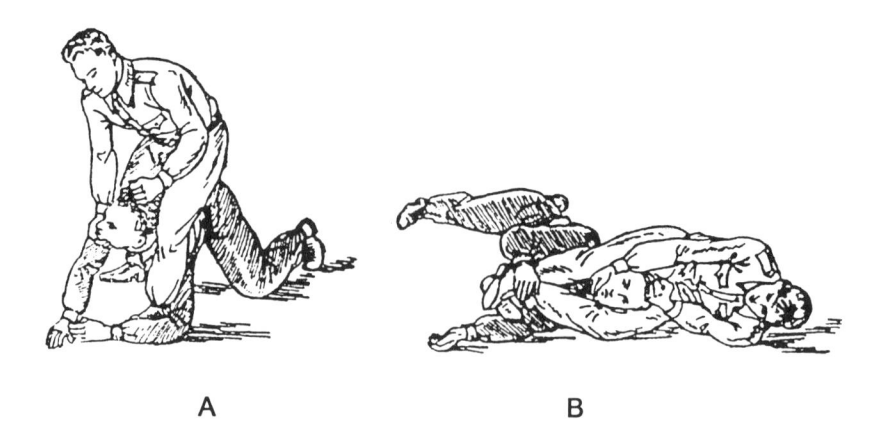

FIGURE 129. CHOKING WITH THE LEGS
A—STRADDLING THE NECK
B—CHOKING, LYING ON THE SIDE

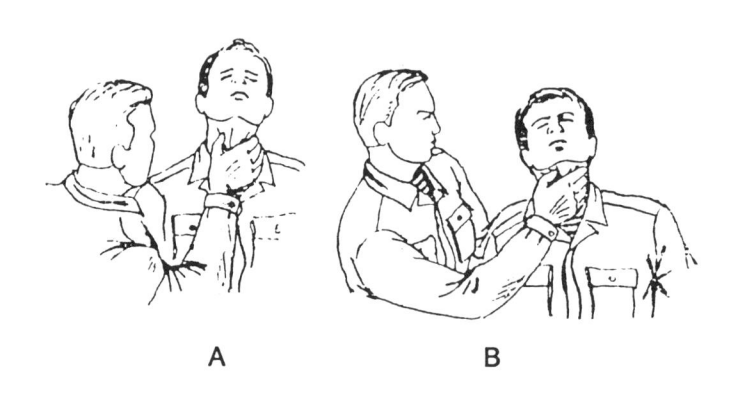

FIGURE 130. CHOKING WITH THE FINGERS
A—FROM THE FRONT
B—FROM THE SIDE

thyroid cartilage region with your fingertips. Use the free hand to hold the enemy's head in place from the back. Press in with the fingertips forward and upward at a 45-degree angle or squeeze the cartilage, jerking it right and left and bending the throat to produce suffocation (Figure 130).

Choking with the Forearm While Pushing on the Back of the Head

This method is fundamental in attacking an enemy from behind. Wrap the right arm around the target's neck, with the elbow snugly under his chin. Put your upper left arm on his left shoulder and lay the straightened fingers of your right hand in the crook of your left elbow (Figure 121). Then put your left forearm and hand to the back of his head or nape of his neck and choke by squeezing your two arms toward each other.

These are additional basic ways to seize and kill an enemy.

FIGURE 131. CHOCKING WITH THE FOREARMS
A—FIRST STAGE
B—SUFFOCATING

ATTACKS BY TEAMS

Silent Attacks on the Enemy from Concealment

A capturing team is operating with four or five members. It selects, say, a communications trench between two firing points as an objective. After moving up unseen to the trench, three men spread out along one edge of the trench, two to one side and one to the other (left and right). To the right and left, farther out, at about three to five paces away, are the fourth and fifth members of the team. All carefully carry out surveillance. When one or two of the enemy come along inside the trench, the whole team jumps down on them to subdue and gag them. The fourth and fifth (flanking) members of the team quietly kill the enemy with knives or garrotes. One member has to be the lookout, keeping his weapon and grenades ready.

ATTACKING AN ENEMY IN ITS POSITION

A trench between two firing joints is chosen for taking an enemy soldier prisoner. The attack team approaches the trench in two groups. Their movements are closely coordinated, based on prior preparation and practicing. Two soldiers from the first group carefully go down into the trench and, simultaneously rising back to back, carefully listen and observe. Two pairs lie at either edge of the trench, not far from each other. One pair lies nearer to a strong point and the other is within four to eight paces from the first. All keep a watch on the strong point. In the same way, the second group positions itself at the other end of the trench.

A few enemy troops appear in the trench. The attackers at the trench's edge let them go by; the pair closer to the strong point slips into the trench to block any escape. Continuing along the trench, the enemy soldiers come upon the blocking pair at the trench's midpoint and are met by cold weapons and/or automatic weapon fire. (Those trying to escape always run in the opposite direction.) The pair at the trench's edge lets the first fugitives run past and shoots the others. The remaining fugitive runs into the arms of the two members of the capturing pair, who have dropped into the trench just for the capture. They knock the fugitive down, disarm him, and take him back along an already chosen route.

In both instances of capturing one enemy soldier, if one or another group/team does not meet up with any hostilities within an hour or two, it can proceed boldly via the trench to one of the strong points. The enemy may take the attackers for some of its own men, and usually will not notice the attackers' passage. Where there is a strong point, a team can fall on the enemy, capturing one and killing the others with knives and other weapons. In each individual case, action is adapted to circumstances, swiftly and audaciously.

In nighttime patrolling and scouting, teams often go out in front as far as five hundred meters. From distances of fifteen to thirty meters, they can throw grenades into enemy dugouts, shelters, and trenches, and then charge into those positions to disarm any left alive.

From cover set up in the enemy's rear, attackers generally have to operate silently, with cold weapons and using attack methods on armed enemies to kill them or render them unconscious for capture.

CAPTURING AN ENEMY TRAVELING BY BICYCLE, MOTORCYCLE, OR HORSE

A capturing team deploys on both sides of a road. Two members of the team lay a rope across the road on the ground, stretching it tight, and concealing themselves whether standing, lying, or sitting, as they hold the rope. The third and fourth members of the team conceal themselves at the side of the road: one carries rope; the other a piece of cloth for use as a gag. The fifth member is the lookout, and he keeps his weapon ready. When an enemy comes along, the first two abruptly raise the rope to the enemy's stomach or chest level and knock him from his seat. The third and fourth members jump on him and secure him while the first two drop the rope and clean up evidence of the capture (and/or catch the horse). After all is accomplished, everyone has to leave the spot quickly.

For coordinating silent night operations, special signals and signs have to be used.

SIGNS AND SIGNALS FOR SILENT OPERATIONS

Hand and arm movements are generally used for signaling in such operations. Headgear, entrenching tools, and other small objects can also be used—but not deviating from established regulation signaling—in situations that preclude voice signals, whistles, shots, or other audible means.

1. To deploy a unit into an open [skirmish] line, put both

arms out to your sides a few times.

2. To make the unit scatter slowly and silently, bend both arms slowly at the elbow and put your fists out to your sides. To make the unit disperse quickly, put your arms quickly out to your sides.

3. To make all lie down, raise a folded arm to chin level and motion downward quickly with the palm, which is facing downward.

4. To get the unit's attention, raise your hand to head height.

5. To have them assemble on you, raise the arm high and make motions above the head and then sharply drop the arm.

6. To indicate movement in a particular direction, raise the hand to head height and then to shoulder height, and then point out the direction.

7. To indicate having spotted an enemy, extend the arm fully out to the side and hold it there.

8. To indicate having heard or seen a signal, raise both hands to head height and then drop them.

9. For a silent approach to an enemy, point the direction out with the left hand and make a few zigzagging motions with the right.

10. For subduing and securing an enemy, make hitting gestures against your jaw or head with left fist, and make two or three circular movements above the fist with your right hand.

11. To indicate specific requirements—knifing an enemy, looking in a window, surrounding a house, climbing a tree—make some clear, expressive gestures with one or two hands.

12. For moving, deploying, and redeploying silently at night, all signaling is done by quick signs learned beforehand or by using light, quick touches on team members' shoulders, chests, backs, or headgear, as well as by taps on the body.

SOME TRAINING EXERCISES AND TASKS

Make sure that a route used in training has plenty of brush, woods, and hills along it. Conceal people along its sides to create surprise situations. A soldier moving along this route should, for example:

A PRACTICAL SECTION IN SPECIAL PHYSICAL TRAINING

- Maneuver past and not touch wires at a height of ten to twenty centimeters that are hidden in the grass, or step on dummy mines
- Move quickly past a rope or pole stretched across the road at chest height by diving or crawling under
- Stop at or run past a falling tree or plank or fend them off with a hay-stuffed backpack
- Free himself and escape from two attacking soldiers
- Disappear and escape from a foxhole

For effective movement in silence in a darkened abode, a trainee should, for example:

- Go from a dimly lit part into complete darkness and move through the completely dark part by touch, not making a sound and finding the door with his hands
- Find a flashlight, light it for an instant, and then find his way out of the dark room in three to five seconds

These tasks can be adapted to different situations.

METHODS FOR SECURING AND TRANSPORTING PRISONERS

Securing is the logical next step after making an enemy prisoner. (Conveying or transporting him with an armed escort

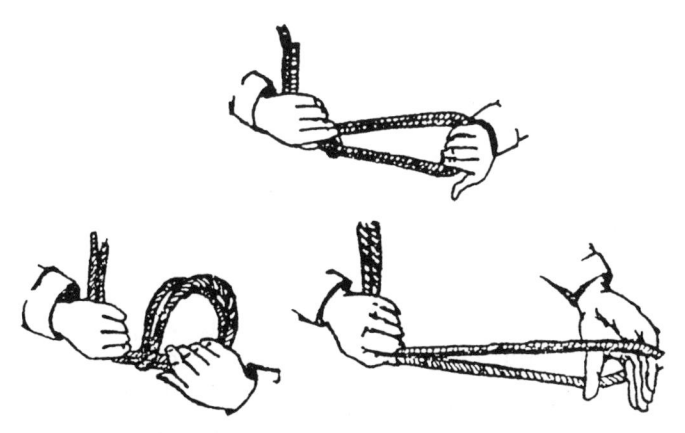

FIGURE 132. THE "LASSO"

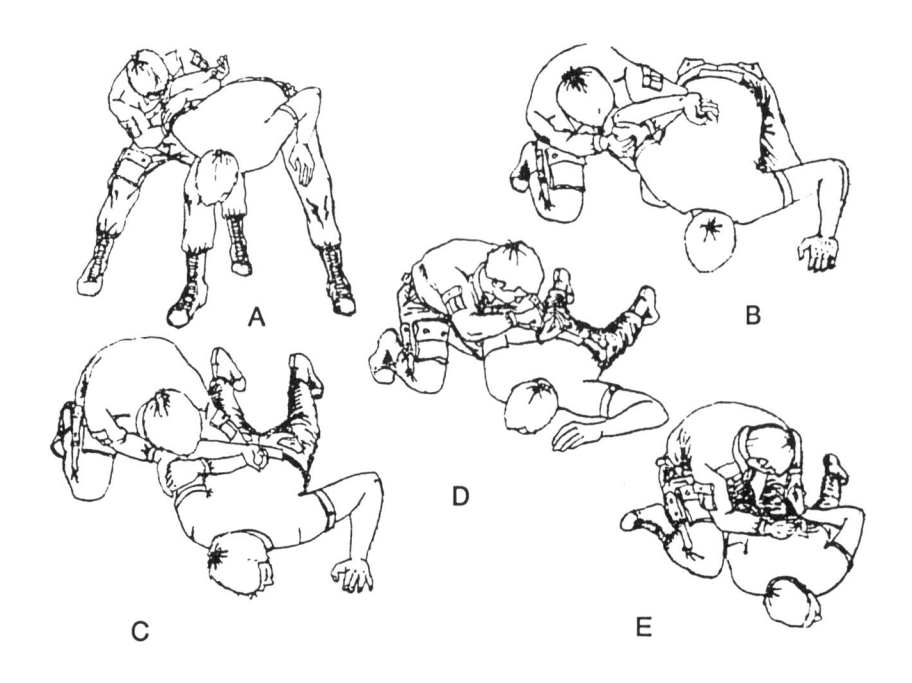

FIGURE 133. SECURING A PRISONER
A,B—TWISTING THE ARM BEHIND THE BACK
C—TAKING DOWN TO THE GROUND
D,E—TYING THE HANDS BEHIND THE BACK

usually follows securing.) He has to be gagged and blindfolded. He has to be searched on the spot and/or hauled off to the side. A prisoner can be tied up with rope, straps, cable, bandages, and so on. For quick tying, a "lasso" can be used. Figure 132 shows how to make this.

First Method of Securing

Twist one arm behind the prisoner's back, forcing him to the ground for tying. The partner twisting the arm keeps it in a lock and exerts painful pressure (by pushing up on the arm) and forces the prisoner to the ground, face down. This same partner gets down on his right knee next to the prisoner to keep him subdued, and loops the rope (lasso) and the hand he is holding, and wraps the rope around the prisoner's wrist a few times. He then ties the prisoner's other hand to the first one and secures both together behind the back with the rope's end (Figure 133).

Second Method of Securing

There are a few variations on the first method, but here, instead of using an arm twist, the prisoner's legs are grabbed from behind and he is thrown down (Figure 134).

After throwing the prisoner, straddle his back and press to the ground. Take one of his hands and tie it. Then pull in his other hand, placing it on top of the first, and bind it too. Tie the free end of the rope around h is neck and secure the end to the tied hands (Figure 135). Lift him up and search his pockets, readying him for further disposition.

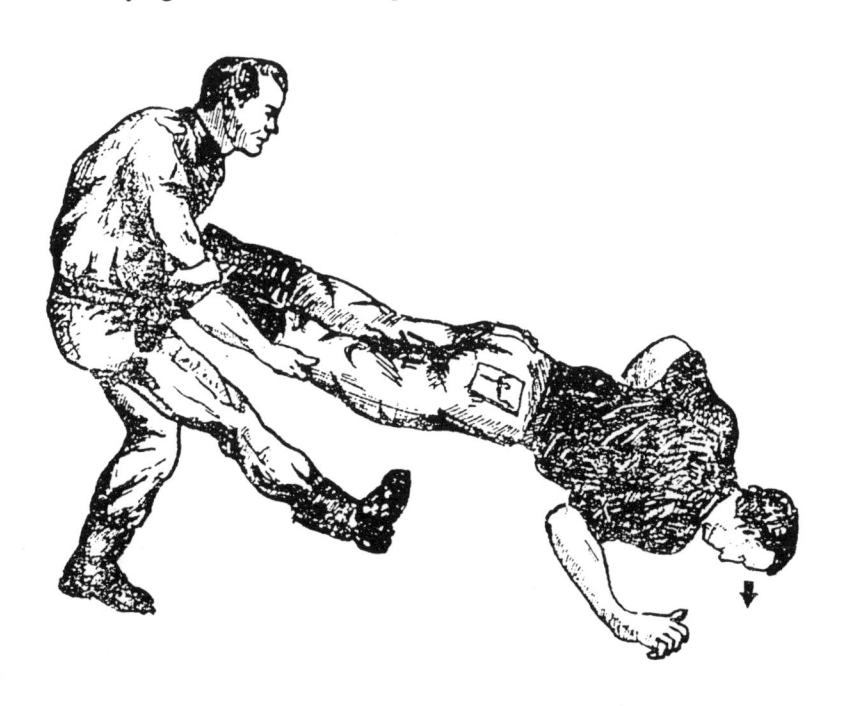

FIGURE 134. THROWING A PRISONER BY THE LEGS FOR SECURING

FIGURE 135. SECURING THE HANDS BEHIND THE BACK

Third Method of Securing

Tying a prisoner's hands together with their palms outward is done after a blow to his head if he offers resistance during capture or tries to escape—or in other circumstances that demand this measure. The prisoner is gagged with a piece of cloth or wood, which is not inserted so deeply as to choke him. To tie his hands, throw the "lasso" around his wrists, tighten the loops and, keeping the ends of the rope together, wrap two to three turns around the wrists, ending up by knotting the whole tie (Figure 136).

It is also possible to take one longer end of the rope and put it around the prisoner's neck and tie the end to the wrist bonds. Seizing from behind for securing is a basic method of effective action against an armed enemy. Speedily seizing and tying a prisoner can also be done from the front. To tie the hands, with palms facing each other or turned outward, the lasso has to be used, drawing it tight and wrapping and tying the ends around the wrists a few times (Figure 137).

In the event that a prisoner will have to cover part of the way by crawling, his forearms can be tied in front with one on top of the other (Figure

FIGURE 136. TYING THE HANDS, USING THE "LASSO"

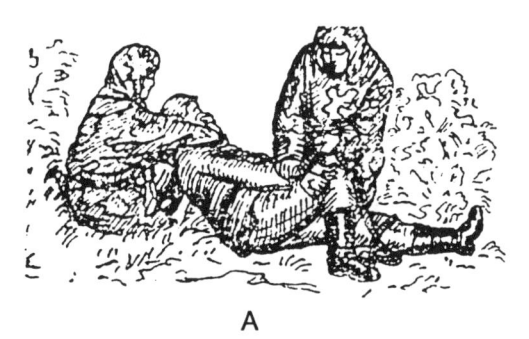

FIGURE 137. TYING THE HANDS IN FRONT
A—PALMS TOGETHER

FIGURE 138. TYING THE FOREARMS TOGETHER

138).

If moving out right after a capture is not possible, and a mission still has to be carried out, tie the prisoner's hands and feet. Tie both hands together and use the longer end of the rope to go around his neck and tie it to the shorter end (Figure 139).

With this, you can hide the tied-up prisoner and leave him for a while. Sometimes a prisoner will be too stunned, and he can be tied by the feet alone. The tying method chosen depends on the disposition (conveyance) planned for hum.

Also, the thumbs can be tied together with cord or rope. For this, the palms have to be together. First, wrap a turn of the rope around the thumbs and run both ends of the

FIGURE 139. TYING THE FEET

rope together between the palms and around the hands once or twice. Tighten and knot. To move the prisoner out, stay by his side or cover him with your weapon.

THE USE OF HANDCUFFS FOR SECURING

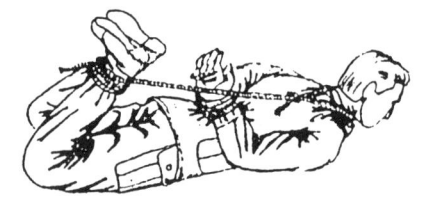

FIGURE 140. HANDS AND FEET TIED BEHIND

In securing a captured enemy, handcuffs can be used instead of rope. They are put on a prisoner who is either standing upright or lying prone.

Standing Upright

Twist the prisoner's arm behind his back. Twist the hand into a painful position and order him to put his left hand behind his head (Figure 141). Take the cuffs in your left hand and put a cuff on his left wrist. Holding on to the cuffs, move his right hand behind his back and snap the other cuff onto that wrist. Convey him as he is.

Lying Prone

Lever the prisoner with his own arm down onto the ground and turn him so that he is face down. Put a cuff onto the wrist of his arm that you are holding. Twist that arm behind his back,

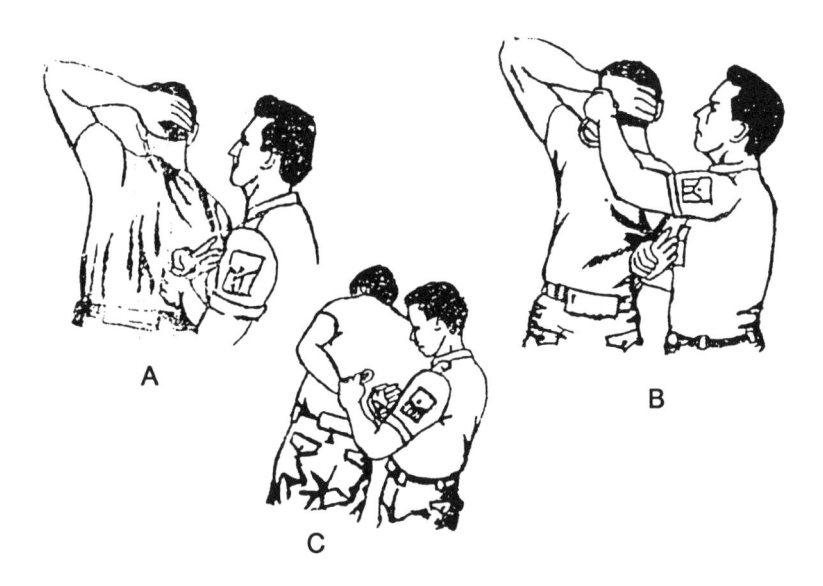

FIGURE 141. HANDCUFFING A STANDING PRISONER
A—TWISTING THE ARM
B—HANDCUFFING THE FREE ARM
C—HANDCUFFING THE HELD ARM

applying painful pressure. Make him put his other arm behind his back and snap the other cuff onto that wrist. Lift him up for further disposition (conveying).

METHODS OF CONVEYING A PRISONER

Methods of conveying and disposing of prisoners can be quite varied, and they are dependent upon things like the area of operation, its condition, proximity of the enemy, and the presence of means of transport.

Conveying (Under Guard)

This method is used when conditions permit walking upright. The prisoner may be bound or not, but he has to be covered with a weapon.

Best and fastest is using handcuffs. Once you put them on his

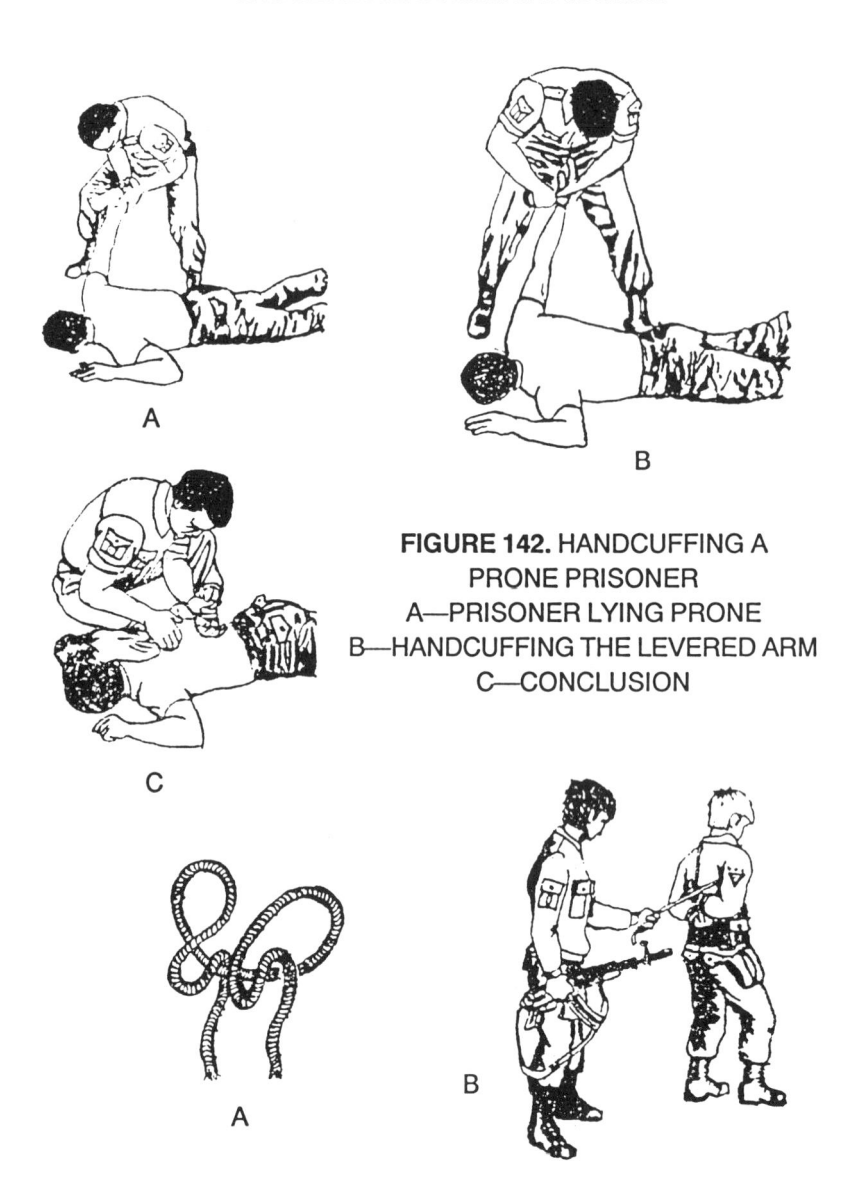

FIGURE 142. HANDCUFFING A PRONE PRISONER
A—PRISONER LYING PRONE
B—HANDCUFFING THE LEVERED ARM
C—CONCLUSION

FIGURE 143. DOUBLE LOOP TIE
A—THE LOOPS
B—CONVEYING THE PRISONER

raised hands and he has lowered his hands, put your left hand on his shoulder and your pistol's muzzle between his shoulder blades. This ensures security. You can make a double loop from a rope or strap, drawing each loop through the other, and secure his hands with this (Figure 143).

Tighten it and keep hold of the free end(s). Stay behind him to the right and keep your weapon in your right hand. Convey him to, e.g., cover or a vehicle. This technique is used in actions in mountain areas.

Transport

This method is used if a prisoner shows signs of resistance when being conveyed away on foot. He can be taken down and have his feet tied to his hands (which are tied in front), and be dragged away on his side (Figure 144).

When hauling away a prisoner by, for instance, his collar, and he is lying on his side, a cap should be put on his head or he

FIGURE 144. DRAGGING A BOUND PRISONER ON HIS SIDE

should be carried in some way so as to avoid injuries.

A variation on transporting a prisoner is to use a rope to drag him on his back (Figure 145) or on a rain cape (Figure 146).

If dragging away a wounded soldier by rope is impossible, then he will have to be carried.

A wounded prisoner can be carried by one soldier on his shoulder or by two soldiers (Figure 147). In either instance the prisoner can be bound. A wounded prisoner can be carried in a

FIGURE 145. DRAGGING A BOUND PRISONER ON HIS BACK WITH A ROPE

FIGURE 146. DRAGGING WITH A RAIN CAPE

FIGURE 147. CARRYING A WOUNDED PRISONER
A—ON THE SHOULDER
B—BY TWO MEN

litter made from two poles and a rain cape or rope. Gagging him is necessary but do it gently.

Carrying a prisoner on a pole can be done using various available gear and other materials (Figure 148).

In mountain operations, prisoners have to be pulled up out of ravines, brought down off precipices, and so on. Therefore, a prisoner should be tied and moved as shown in Figures 149 and 150.

A PRACTICAL SECTION IN SPECIAL PHYSICAL TRAINING

FIGURE 148. CARRYING A PRISONER ON A POLE

Recommendations

After capturing a prisoner with his papers and weaponry, all traces of the attack should be removed: the enemy may very well discover signs of your action later. A swift withdrawal and conveying of the prisoner must be organized, with

FIGURE 149. HAULING A PRISONER OUT OF A RAVINE
A—BY THE COLLAR
B—WITH A ROPE

FIGURE 150. BRINGING A PRISONER OFF A HEIGHT BY ROPE

security and cover for the capturing team. In the event of a sudden encounter with the enemy, attack first, using all firepower at hand.

METHODS FOR EVACUATING THE WOUNDED

In combat, it is very important to evacuate the wounded from the battlefield to cover or a safe place before rendering medical help.

In line with this, immediately after mastering the various ways of crawling, the following exercises should be practiced: moving a comrade by crawling, carrying a comrade single-handedly, carrying a comrade over a great distance (individually or with two persons).

First Exercise

In evacuating a comrade by crawling, clamp his upper arm to your body (Figure 151). Crawling like this is done low to the ground, using elbows and feet. There is the method of carrying a wounded comrade by crawling and holding his arm over your shoulder (Figure 152).

FIGURE 151. CRAWLING WITH A PARTNER, HIS ARM CLAMPED UNDER YOURS

Second Exercise

For moving forward on your right side (Figure 153a) or on your back (153b), grasp your wounded comrade's collar or put a belt or strap under his upper arm. Move forward on your side or back, steadily working with your elbows and feet. During all this,

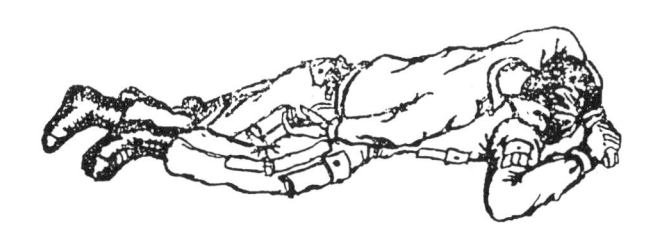

FIGURE 152. CRAWLING WITH A PARTNER, HIS ARM OVER YOUR SHOULDER

A

B

FIGURE 153. EVACUATING A PARTNER
A—ON YOUR SIDE
B—ON YOUR BACK

his head must rest to the back and be raised off the ground. In a safe area nearby, carrying a comrade can be done more quickly.

Third Exercise

This is done by transporting the comrade on a rain cape, using a rope to pull it (Figure 154). The comrade has to be

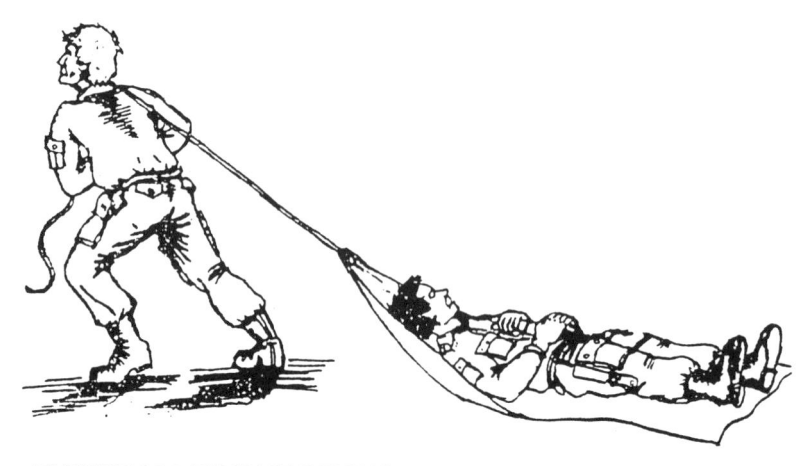

FIGURE 154. TRANSPORTING A PARTNER ON A RAIN CAPE

secured to the cape and has to lie on his back.

Fourth Exercise

Carrying a comrade is done by:
• a shoulder carry ("grain sack") (Figure 155a)
• a piggyback carry with your arms under his knees (Figure 155b)

These two methods are used in the event of wounds to the back.

Group methods of carrying also have their variations (Figure 156).

The assumption is that these methods are to be used for relatively short distances; longer-distance transport is by armored vehicle or other kinds [of transportation].

The training exercises can be done on streets or in gymnasiums, in exercises where crawling is important and where the trainees will need a lot of physical development.

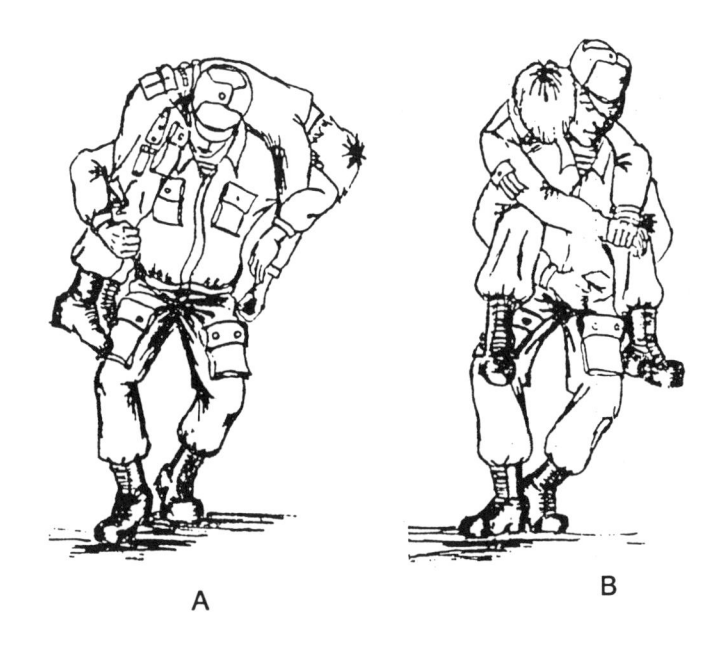

FIGURE 155. CARRYING A PARTNER SINGLE-HANDEDLY
A—OVER THE SHOULDERS
B—ON THE BACK

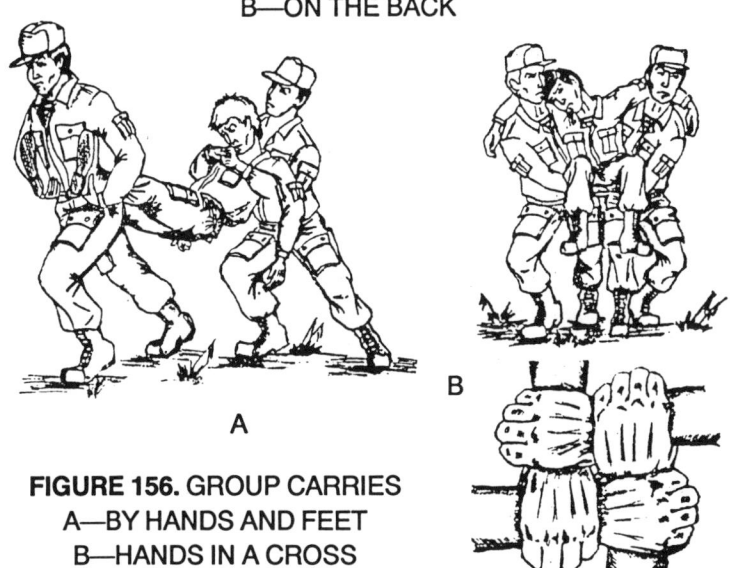

FIGURE 156. GROUP CARRIES
A—BY HANDS AND FEET
B—HANDS IN A CROSS

CHAPTER 5

ESCAPING FROM AND FIGHTING OFF PHYSICAL ATTACKS; MUTUAL AID; THROWS

It is very important to learn not only how to attack, but also how to defend yourself from an attack. You use a firearm, rifle butt, or bayonet; and employ the back of your head, your feet, your hands, and many other means. Prearranged signs and signals have to be used in any sudden attack on an enemy, under any conditions; and avoiding capture has to be a continual concern. In all this, killing or capturing the enemy has to be a goal.

For example, if an enemy grabs you from behind or the side, by the neck or the arms, you have to smash your head into his face and kick him in the knee hard (Figure 157). Once free, you use your fists, cold weapons, or a rifle. If an enemy throws an article of clothing over your head to try to take you prisoner, you should thrust your arms up vigorously and thus escape it. For a successful attempt to capture an enemy, you seize him by the feet and throw him to the ground, and then you use a chokehold—or a knife or bayonet to kill him. Throwing an enemy by grabbing his legs from the front is done after you have first hit him in the nose or eyes (if he has tried to capture you) (Figure 158).

Throws and other hand-to-hand techniques are the best and most successful methods of breaking free of capture attempts. These are not done just as academic exercises. A soldier in full field gear and carrying a weapon should learn how to perform these methods simply and boldly without worrying about an

FIGURE 157. ESCAPING A GRAB FROM BEHIND

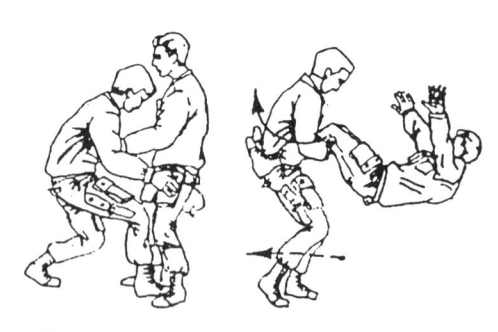

FIGURE 158. THROW, GRABBING THE LEGS FROM IN FRONT

FIGURE 159. ESCAPE BY BANGING AN ENEMY'S HEAD AGAINST A WALL

FIGURE 160. ESCAPE BY A BLOW TO THE FACE AND TRIPPING

ESCAPING FROM AND FIGHTING OFF PHYSICAL ATTACKS; MUTUAL AID; THROWS

FIGURE 161. ESCAPE BY TRIPPING AN ENEMY AND KNOCKING HIS HEAD ON A WALL

FIGURE 162. POSTURE FOR KICKING WHILE ON YOUR BACK

enemy's safety, but instead making the attack more forcefully (Figures 159-161).

In escaping being grabbed, you should, above all, utilize the most important vulnerable points of an enemy's body: the head, the ears, the throat, the sex organs, and so on. If an enemy gets you down on the ground, you have to continue the struggle by using your knee, by choking, biting, gouging, and kicking (Figure 162).

The following material will be presented:
• Escaping and fighting off attacks from behind
• Escaping and fighting in single combat when down on the ground
• Giving mutual aid to escape or capture attempt
• Throws and other techniques with variations on related training in escape methods

ESCAPING ATTACKS FROM THE FRONT

If an attempt made to

193

capture you isn't very sudden, you can repel it with blows to a foe's vulnerable spots, such as with fingers to his eyes or a kick to the crotch (Figure 163), or by using objects at hand, such as a briefcase or attaché case (Figure 164).

If an enemy grabs your shoulders, chest, throat, or arms, you should use the fundamental butt with your head in his face or fingers in the eyes (Figure 165), after which you use the throw where you grab his legs from the front, or a hit to the bottom of his nose, or tripping (Figures 160 and 161). After a throw, put a knee in his crotch (Figure 166).

The escape techniques presented earlier (choking, twisting, hitting) are the primary means of successfully breaking free and performing the final moves. The most dynamic technique is the blow. To escape restraints, grasps, and holds, and to facilitate using your weapon, use these blows:

• Fingers (index and middle) in the eyes, done with a short jab quickly withdrawn and followed with shooting or using a cold weapon (Figure 165)

• Head butt in the face if an enemy has you by the arms, shoulders, or belt, done directly to the nose

• Hand/arm blows using the fist, edge of the hand, or elbow, swung from below into the stomach, neck, or face (Figure 168), and other spots, depending on circumstances

FIGURE 163. STOMACH KICK

FIGURE 164. ATTACHÉ CASE IN THE CROTCH

FIGURE 165. FINGERS IN THE EYES

FIGURE 166. KNEES IN THE CROTCH AFTER A THROW

FIGURE 167. HEAD BUTT IN FACE

To make your escape complete, it is recommended that you get your foe down and kill him.

Aside from these blows used against an attacker from the front, you should attack other points of the body, such as up under the nose, under the ears, the eyeballs, as well as the lower lip and the muscles at the side of the neck (Figure 169).

For escape, crushing, twisting, and other pain-producing actions are combined with kicks to, e.g., the shin, crotch, or stomach. These produce extreme pain that incapacitates an enemy and can produce wider damage.

Against an attack at the legs from the front, the following can be used:
 • throws over the back (but for this there is a special fall that is used in practice and is quite dangerous!)
 • knees to the face and chest and an elbow in the back

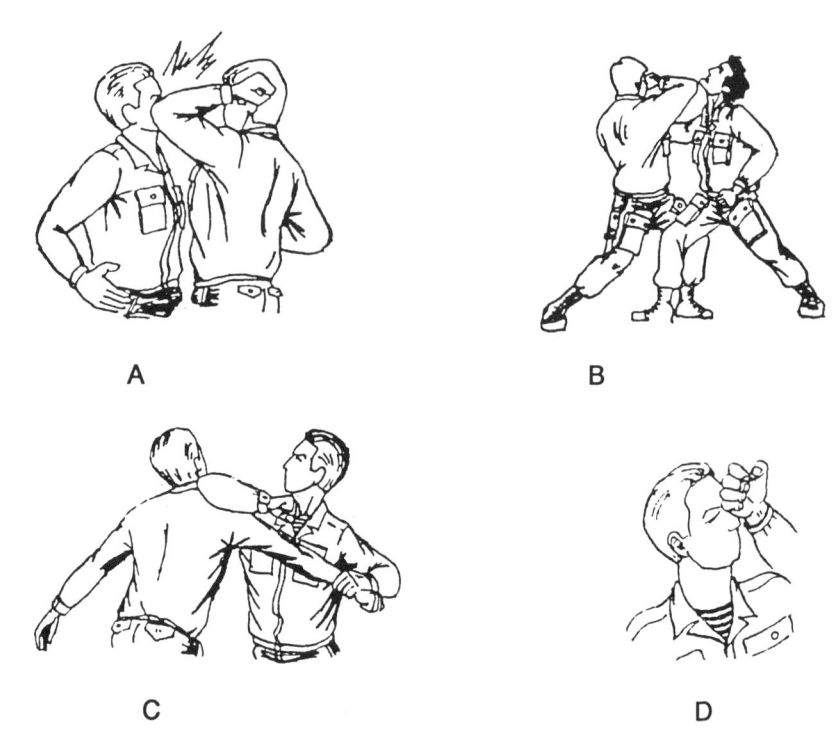

FIGURE 168. BLOWS WITH FIST AND ELBOW
A—IN THE FACE
B—TO THE CHIN
C—FROM THE SIDE
D—ON THE NOSE BRIDGE

between the shoulder blades against the vertebrae (very effective, Figure 170)
• clamping the head for choking

Choking is done with an arm around the throat with a quick, upward pressure, squeezing or twisting the head left and right (Figure 171). The best way to escape from a frontal attack is to respond by grabbing the assailant's crotch. This requires a forceful grasping and twisting back and forth, and trying to pull your attacker toward you. The method is simple but effective.

ESCAPING FROM AND FIGHTING OFF PHYSICAL ATTACKS; MUTUAL AID; THROWS

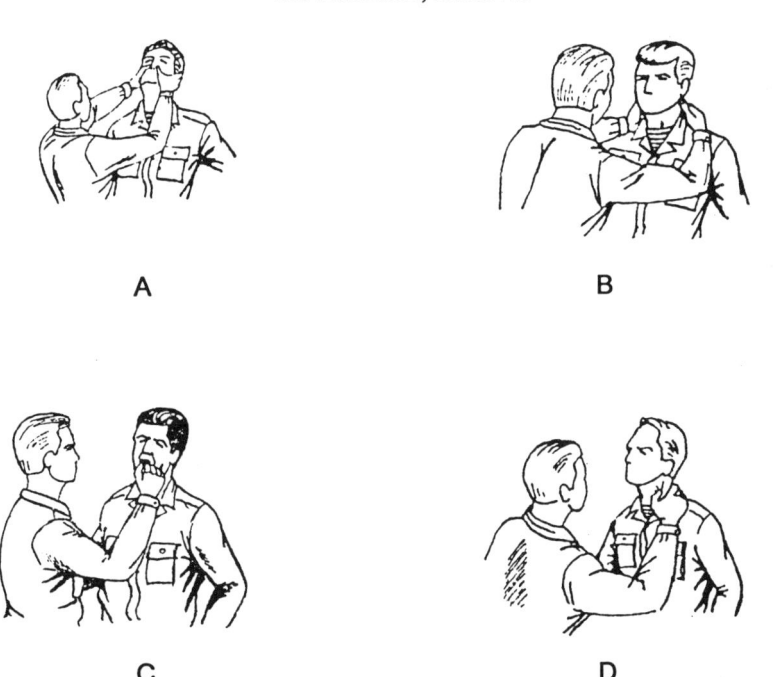

FIGURE 169. OTHER ATTACKS TO THE HEAD AREA
A—THE BASE OF THE NOSE
B—UNDER THE EARS
C—THE LOWER LIP
D—MUSCLES AT THE SIDE OF THE NECK

Blows to the face and kicks to different vulnerable points are often used in escaping from frontal attacks (Figures 172 and 173).

Additional Ways to Escape from Attacks

These methods also produce damage when used to inflict pain, even though they are more "humane." They include such actions as twisting, knocking down, grabbing the sex organs, blows, and other painful ways of direct contact. Most often, a blows or arm twist is used (Figures 174 and 175).

Figure 176 shows a simple way of getting out of a chokehold: an upward blow with both hands breaks the hold and

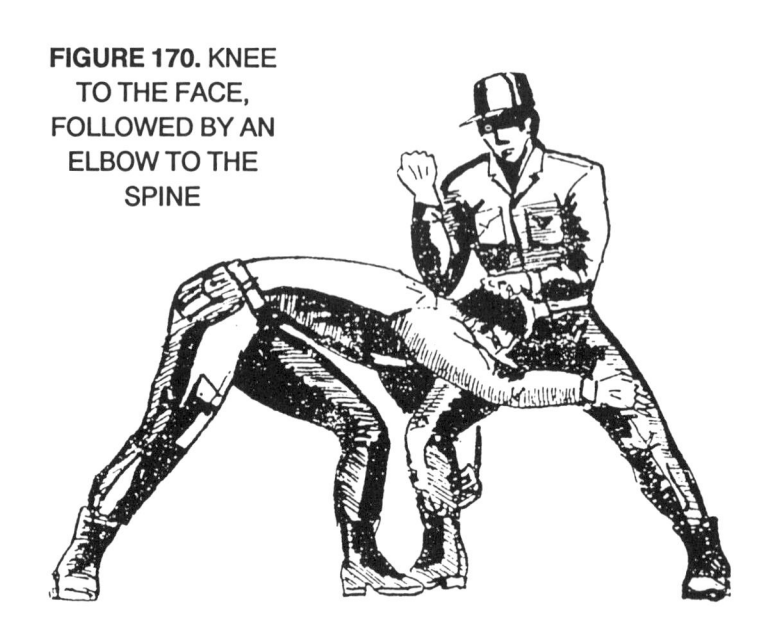

FIGURE 170. KNEE TO THE FACE, FOLLOWED BY AN ELBOW TO THE SPINE

FIGURE 171. CHOKING UPWARD WITH THE FOREARM

ESCAPING FROM AND FIGHTING OFF PHYSICAL ATTACKS;
MUTUAL AID; THROWS

FIGURE 172. KICKS
A—WITH THE KNEE
B—FOOT TO THE CROTCH
C—FOOT TO THE SIDE

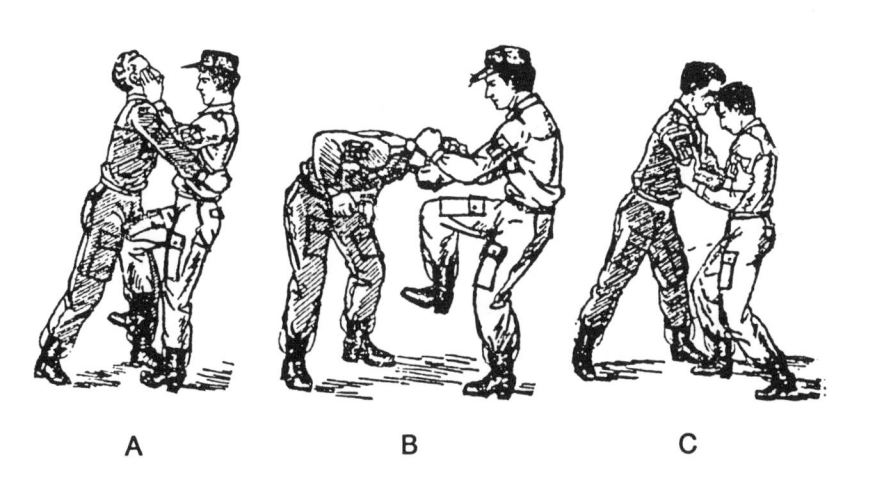

FIGURE 173. COMBINATION OF BLOWS (TO THE FACE)
A—FINGERS IN FACE, KNEE TO CROTCH
B—FACE ONTO KNEE
C—HEAD BUTT TO FACE

KGB ALPHA TEAM TRAINING MANUAL

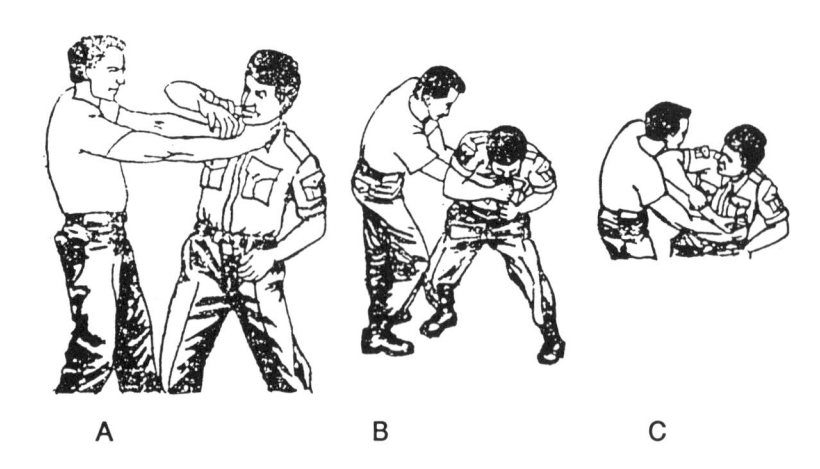

A   B   C

FIGURE 174. ESCAPING A CHOKEHOLD FROM THE FRONT
A—SEIZING THE HANDS
B—GETTING FREE OF THE HOLD
C—ELBOW BLOW TO THE SIDE OF THE NECK

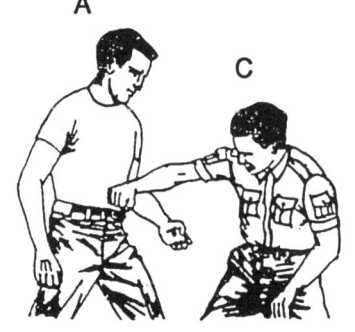

FIGURE 175. ESCAPING AN ARM GRIP
A—THE GRIP
B—ESCAPE
C—PUNCH TO THE STOMACH

ESCAPING FROM AND FIGHTING OFF PHYSICAL ATTACKS; MUTUAL AID; THROWS

positions the fists for a blow to the assailant's face. This blow can be combined with a kick to another vulnerable point.

Figure 177 shows how to break free of a body hold from the front. Butt the assailant in the face with your head or strike him

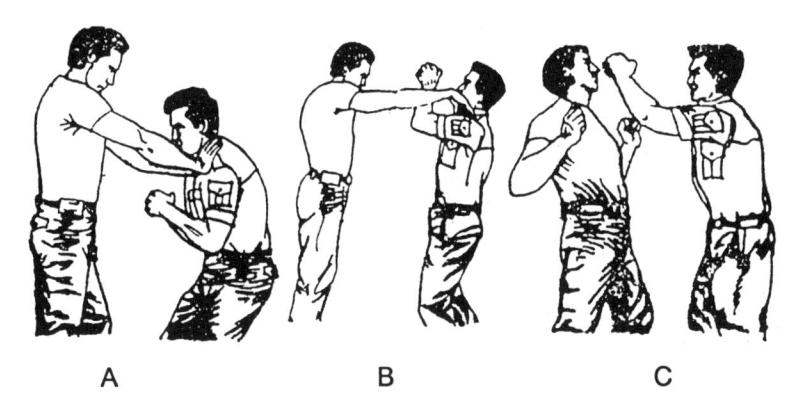

A   **B**   **C**

FIGURE 176. ESCAPE FROM A CHOKEHOLD FROM THE FRONT
A,B—KNOCKING THE ARMS APART
C—BLOW TO THE FACE

under the chin or nose with the heel of your hand. Follow up with a knee to his crotch or a shove under his chin. Once he is on the ground, use a wrap-up kick to his head, ribs, back, stomach, or crotch with either the toes or sole of your foot.

If you need to detain and/or question an assailant after you have broken free, twist his arm back and under as shown in Figure 178. Because this action is not all that simple, it is very important that you understand the following. To effectively escape the hold and then detain your assailant, you have to start with an immediate and powerful kick to his crotch or shin, and then twist his arm. In learning all these methods and techniques for escaping holds, you and your partner must be very careful!

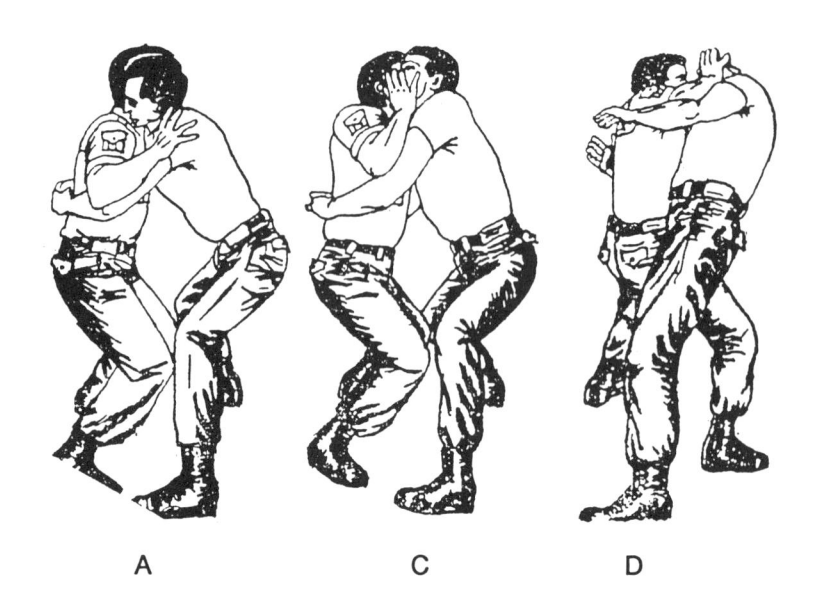

FIGURE 177. ESCAPE FROM A BODY
HOLD FROM THE FRONT
A—POSITIONING
B,C—SHOVING THE CHIN WITH BOTH HANDS

ESCAPING ATTACKS FROM BEHIND

It is exceptionally important to learn and know how to deal with attacks from behind. As a rule, an attack from behind is meant for your capture. You have to resist in any such situation, even if an assailant gets you in a chokehold (Figure 179). As an example, if it proves impossible to throw him over your shoulder—which, given the chokehold, could lead to worse effects on your neck—you can try blows with your elbow or edge of your hand (once you are free of the hold) and moves sideward. If these actions do not seem feasible immediately after you break free of the hold, then other measures should be taken.

The most effective blows are with the elbows, fists, and edges of your hands. Although kicks of all kinds are very effective when your hands and arms are constrained, the arm and

ESCAPING FROM AND FIGHTING OFF PHYSICAL ATTACKS; MUTUAL AID; THROWS

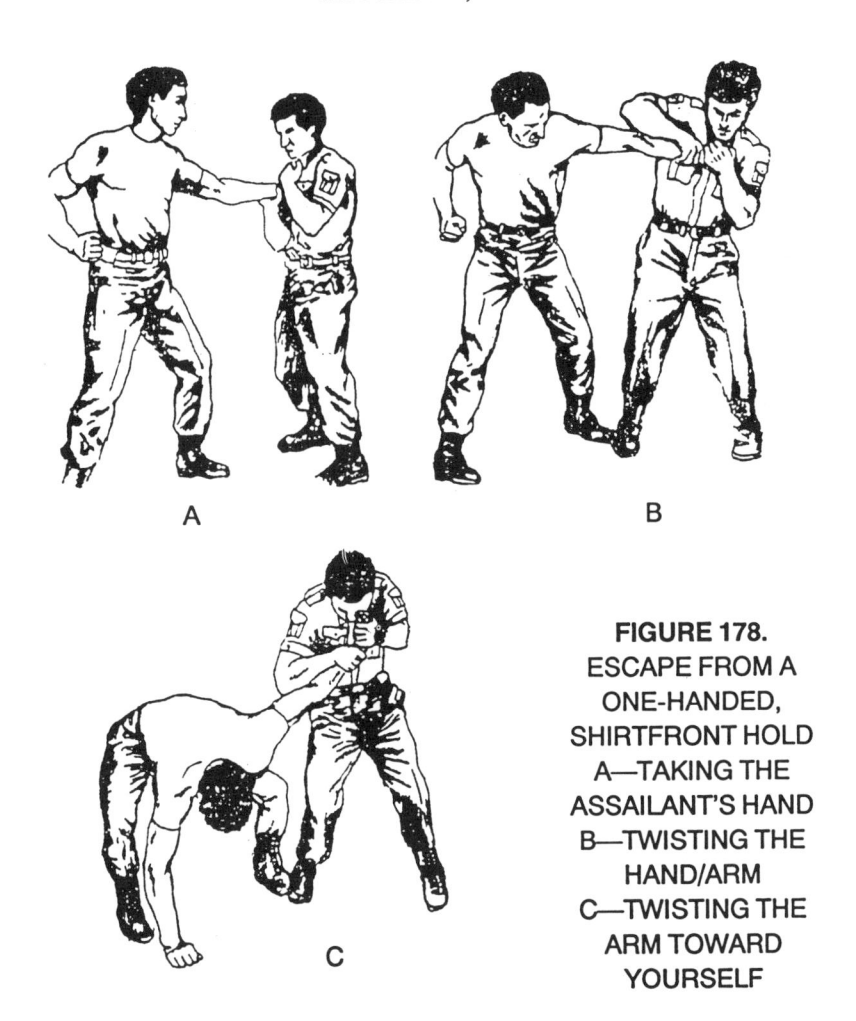

FIGURE 178. ESCAPE FROM A ONE-HANDED, SHIRTFRONT HOLD
A—TAKING THE ASSAILANT'S HAND
B—TWISTING THE HAND/ARM
C—TWISTING THE ARM TOWARD YOURSELF

hand blows are more accurate. Where you aim your blows depends on the situation. If an attacker is not holding your arms below the elbow, you can use the elbows on his midriff, face, or other spots (Figure 180).

Immediately after you deliver a few sharp blows, you have to step away from your assailant, turn to face him, deploy a weapon, and act accordingly.

In attacks aimed at the upper part of your body (hair, neck,

**FIGURE 179.
CHOKEHOLD FROM
BEHIND**

clothes, shoulders, and so on), hitting your attacker in the crotch with the edge of your hand is the best for your escaping the assault (Figure 181). When practicing any of these blows with a partner, simply pretend to hit; at a punching bag, use full force.

Let us now look at more complicated situations and escape methods that also are based on using blows.

Figure 182 shows how to hit with your head when an assailant is using a chokehold or has got you around the arms and trunk—and is pressed up against your back. You should butt his

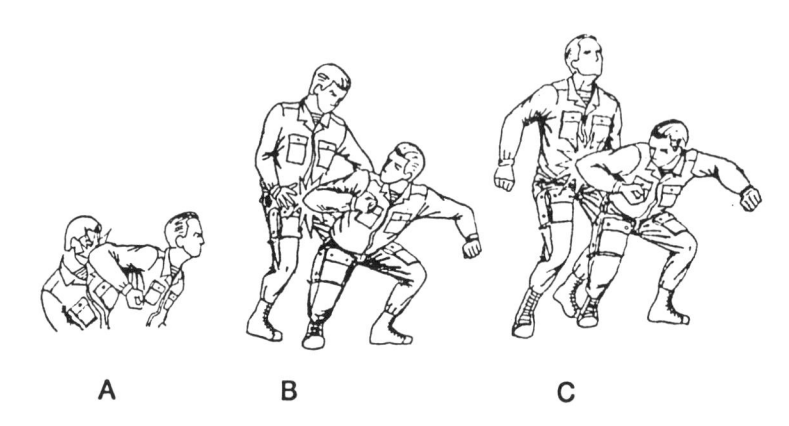

A B C

FIGURE 180. BLOWS WITH THE ELBOW
A—IN THE FACE
B—IN THE CROTCH
C—IN THE STOMACH

face with the back of your head or grab his arms and get ready to throw him. You can throw him forward (by grasping him by his armpit) or backward onto his back by grabbing his lower legs (from between your legs).

ESCAPING FROM AND FIGHTING OFF PHYSICAL ATTACKS; MUTUAL AID; THROWS

FIGURE 181. BLOW TO THE CROTCH

A

B

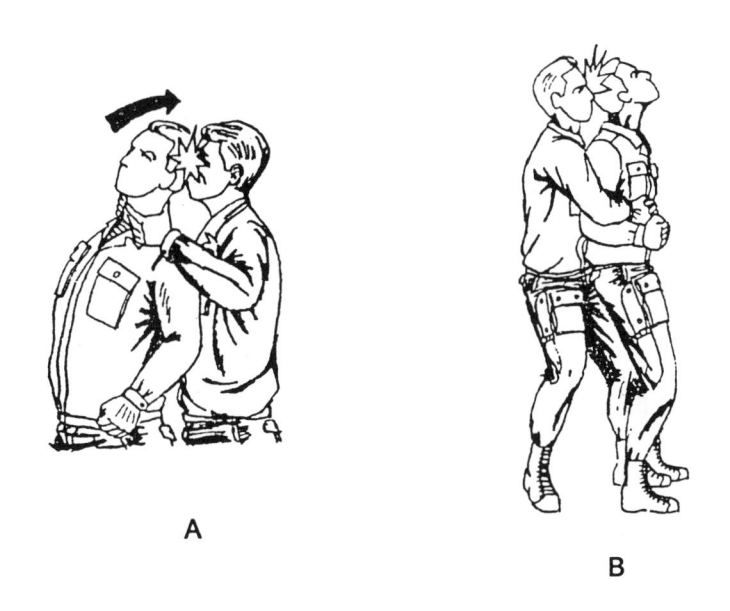

FIGURE 182. ESCAPE FROM A BEAR HUG FROM BEHIND
A—IN A CHOKEHOLD
B—IN A BEAR HUG

KGB ALPHA TEAM TRAINING MANUAL

A B

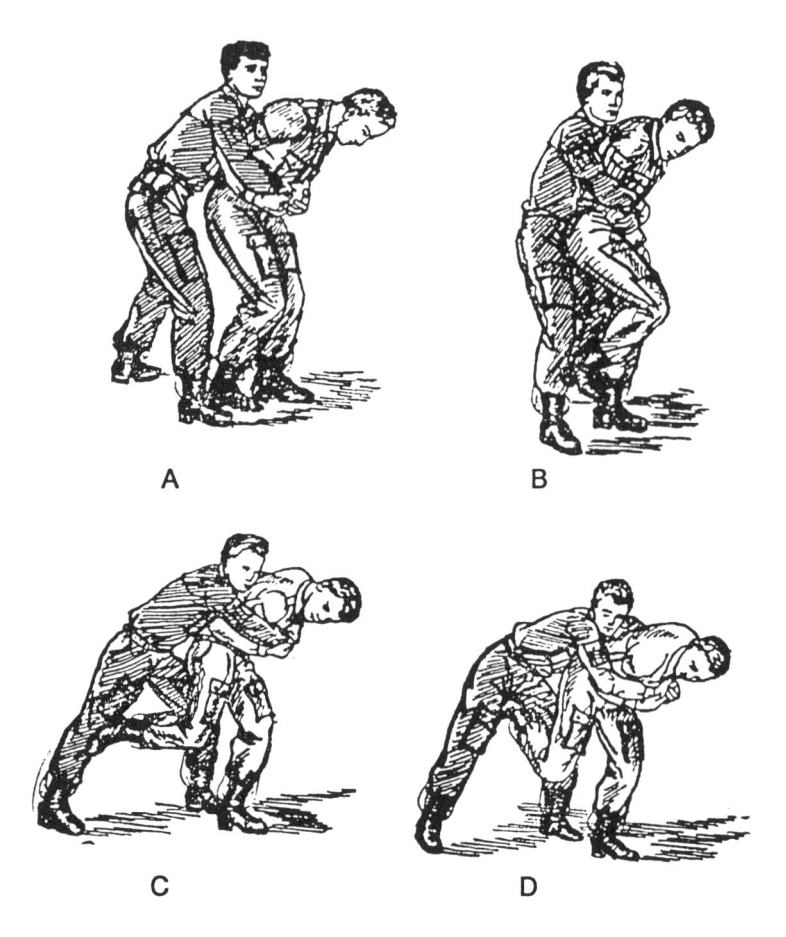

C D

FIGURE 183. ESCAPE FROM A BEAR HUG FROM BEHIND
A,B—UPWARD/BACKWARD KICK
C—KICK TO THE KNEE
D—KICK TO THE CROTCH

Use kicks when the assailant uses a bear hug from behind, pinning your arms to your body. A flat-footed kick, especially if the boot soles and heels have hobnails and horseshoe taps, is awfully effective and painful—and can produce a fracture. First, kick upward and back to force the attacker away from you (Figure 183).

ESCAPING FROM AND FIGHTING OFF PHYSICAL ATTACKS;
MUTUAL AID; THROWS

When your assailant has moved back a bit, follow up with a kick to his shin or knee or up into the crotch, according to circumstances and his position (Figure 183).

After performing all this, have a weapon ready.

Blow Combinations

When an assailant holds you tightly from the rear, use an upward kick to the rear and a blow to his stomach with your elbow (Figure 184).

Neck/head holds are dangerous. You have to break loose quickly or suffer serious suffocation. You have to move your legs

FIGURE 184. ELBOW BLOW TO THE STOMACH AND THROW OVER THE BACK

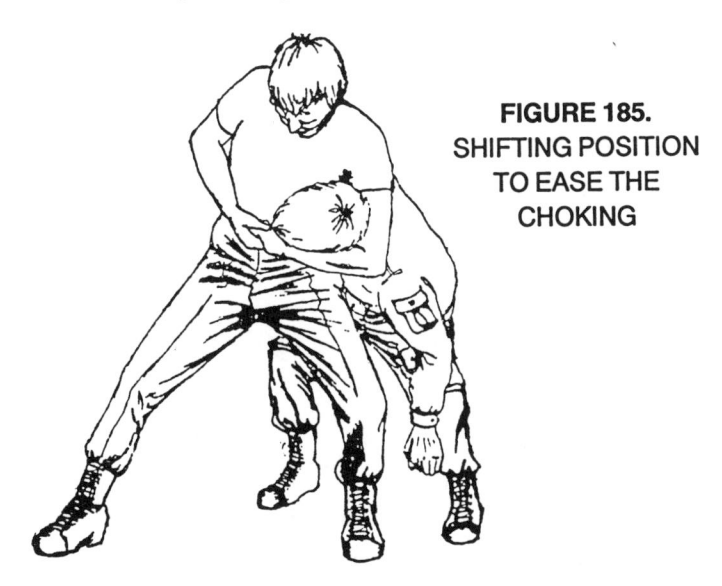

FIGURE 185. SHIFTING POSITION TO EASE THE CHOKING

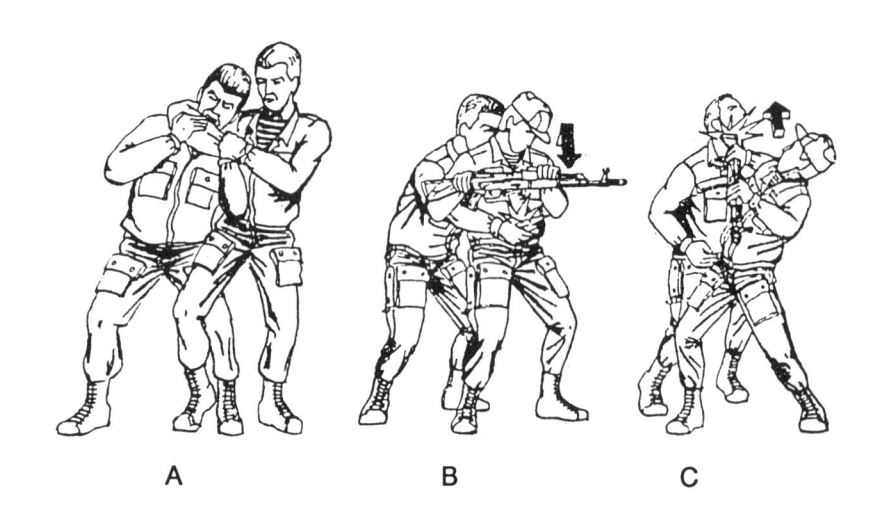

FIGURE 186. BREAKING OUT OF A HOLD FROM BEHIND
A—OUT OF A NECK HOLD
B,C—OUT OF A BODY HOLD

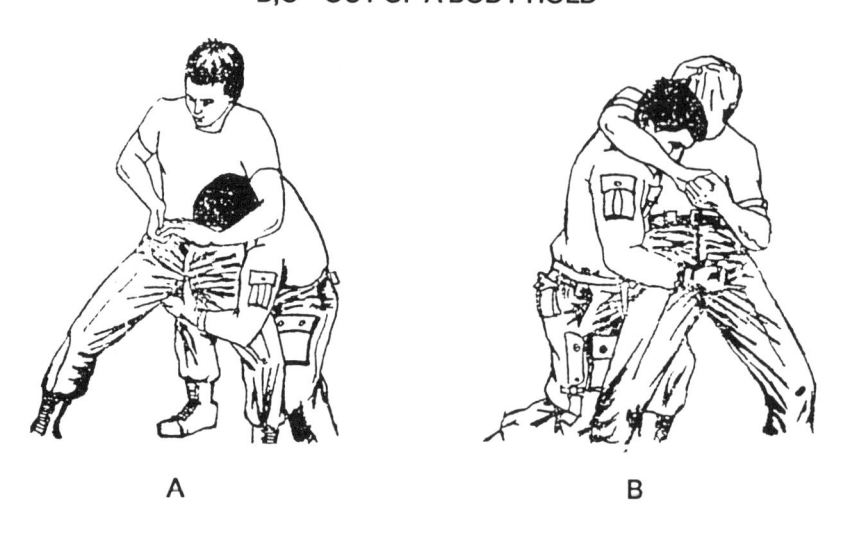

FIGURE 187. ACTIONS AGAINST THE GROIN
A—SQUEEZING
B—PUNCHING

ESCAPING FROM AND FIGHTING OFF PHYSICAL ATTACKS; MUTUAL AID; THROWS

surreptitiously behind your assailant's, as shown in Figure 185. Since you cannot remain in this fix for long, you can use an elbow blow to his stomach or groin and sharply push him back so that he falls on his back, you falling with him. Once on the ground, get completely out of the hold and on your feet. Then take appropriate action.

Simpler yet, get into position to free yourself by going for an assailant's testicles by squeezing or punching them (Figure 187). Use whichever method fits the situation. If it is possible, get your hand to his face and jab his eyes.

Use a combination of attacks to the eyes and crotch to escape.

Basic Rules and Recommendations for Learning the Methods of Escaping from Holds

If an assailant is trying to choke you, you absolutely have to disrupt this with your own hands; that is, you have to loosen his grip on your throat or neck and then be ready with a way to completely free yourself (Figure 188).

You can break deadly holds by sharply twisting an attacker's thumb (Figure 189).

A   B

FIGURE 188. ESCAPING A HOLD FROM BEHIND
A—LOOSENING THE HOLD
B—ELBOW BLOW AND THROW

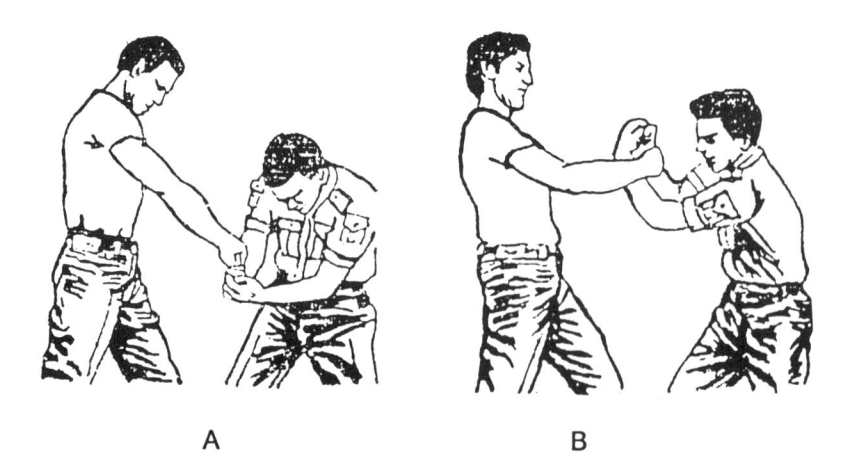

FIGURE 189. BREAKING OUT OF A HOLD
A,B—TWISTING THE THUMB

ESCAPING FROM HOLDS IN FIGHTS ON THE GROUND

Very often, an unforeseen encounter with an enemy will not allow you to use a weapon. Brief attacks can turn into fierce one-to-one fights where all means must be used to win.

An Assailant on Top Tries to Beat You While You Are Supine

Get your attacker in an overarm head lock with your forearm pressing against his throat, put your legs around his body with a scissors hold, and roll over so that he ends up on his back or sitting. Squeeze his body with your legs and bend his head as much as possible, tightening your legs and pulling his head outward from his shoulders (Figure 190).

FIGURE 190. BREAKING FREE BY USING BOTH LEGS

FIGURE 191. BREAKING FREE BY WRENCHING THE NECK VERTEBRAE

An Assailant Gets You Down and Is on Top

Get your legs around the attacker's body in a scissors hold and push up on his chin with your right hand. With your left hand, grab the hair on the back of his head and pull it toward you, wrenching the neck vertebrae (Figure 191).

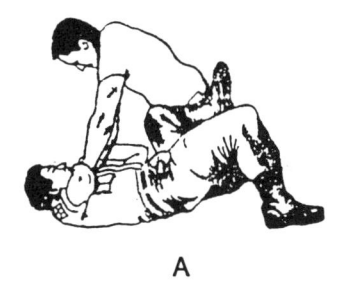

FIGURE 192. BREAKING FREE BY GETTING THE ARM BETWEEN THE LEGS
A—GRIPPING HANDS, RAISING LEG
B—ARM BETWEEN LEGS

An Assailant on Top of or Beside You Tries to Choke You

Escape Technique 1
Take a good grip on his hands and, at the same time, get the knee nearest him under him and between his arms. Push him away from you and, straightening your legs, get his extended arm between them (Figure 192).

Escape Technique 2
Break the attacker's hold from inside, between his arms and simultaneously roll sharply to the side, getting on top of him. In this position, use a chokehold or a fist on him (Figure 193).

KGB ALPHA TEAM TRAINING MANUAL

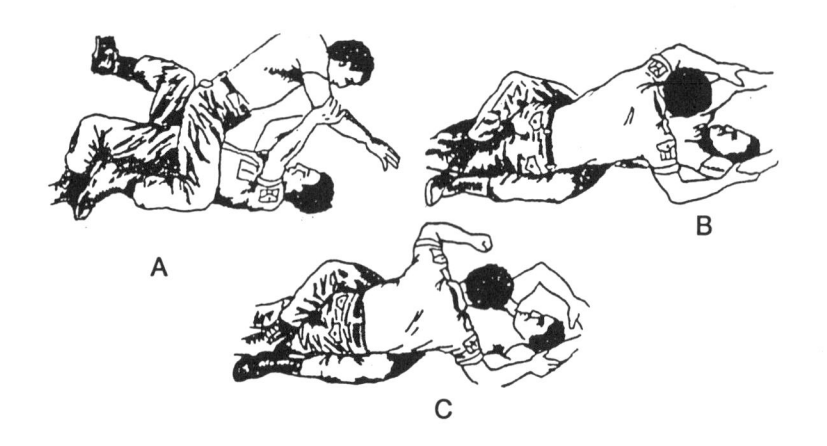

FIGURE 193. ESCAPING AN ASSAILANT ON TOP
A—GRABBING THE UPPER ARMS
B—ROLLING OVER
C—PUNCHING THE FACE

FIGURE 194. ESCAPING A CHOKE HOLD FROM BEHIND
A—GRABBING THE HANDS, ON ONE KNEE
B—THROWING

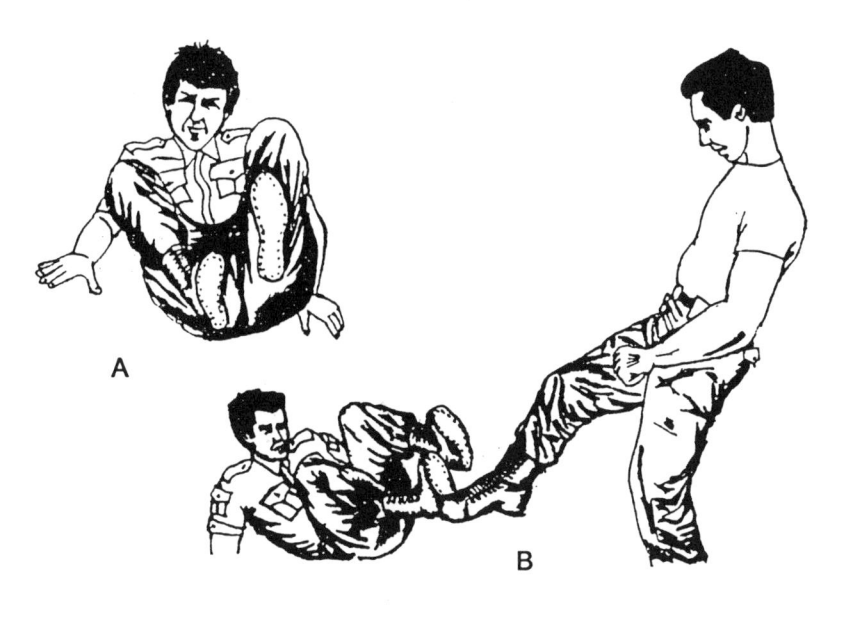

FIGURE 195. DEFENSE POSITION WHEN DOWN
A—POSITION
B—DEFENSE AGAINST KICKS

An Assailant Gets a Chokehold on You from Behind

Quickly drop to one knee, grip the attacker's hands, and bend forward sharply at your waist, throwing him. Put a few blows or kicks into your stretched-out assailant (Figure 194).

An Attacker Has You in a Bear Hug from the Front

Use your hands on his face: the chin, up under the nose, the eyes. Once you get him backed away, trip him and hit him. If you fall down or go to the ground, keep facing your attacker as he approaches you. Gather yourself (if you are on your back) and meet him with heel kicks to his shins or knees. Use the soles/heels of your footgear to ward off his kicks (Figure 195). Staying down too long is dangerous.

If you fall down in a fight, do everything to get back on your feet. The method is shown in Figure 196.

A   B

FIGURE 196. METHOD FOR GETTING UP AFTER A FALL
A—ON ALL FOURS
B—ON BOTH FEET

Get off your back onto all fours, then from your knee(s) to an upright stance. Immediately face your enemy, carefully watching him. In getting up, you may have to throw yourself at your assailant, but it is better to back off, get a weapon, and cover him—then take appropriate action.

DEFENSE AND MUTUAL AID

Blows—the Methods for Help and Rescue

In hand-to-hand fights, when it is impossible is use weapon fire, working out ways to secure a comrade's safety definitely has a bearing on decisive and successful action. "You may fall that your comrade may live." [*"Sam pogibai, no tovarishcha vruchai!"*] This saying implies previous training, however, so that you do not die and that you do save your comrade in a critical situation. It is practically impossible to fire your weapon to help a comrade who is in a hand-to-hand struggle. The basic recommendation for rescue is to get in direct contact with an enemy assailant: attack, using fingers, fists, a knife, or whatever. But when an attacker is choking your comrade, you cannot shove or grapple with the assailant. What is needed right away are kicks, hand blows, or some kind of weapon.

ESCAPING FROM AND FIGHTING OFF PHYSICAL ATTACKS;
MUTUAL AID; THROWS

FIGURE 197. KICK TO THE HEAD TO HELP

We will examine some instances when this kind of help will be urgently needed. Say an enemy has gotten on top and is pressing down on a comrade's throat with a rifle. The rescue method is simple: kick the attacker quickly and powerfully in the head; then take your rifle and take him out with a butt stroke or put it across his

FIGURE 198. A KICK TO THE FACE

throat. As an alternative, run at him from the back or side so that he does not see you. For similar results, give him a kick in the face, side, back, or tailbone. By and large, a kick in the face is recommended because he may be wearing a flak jacket, which will protect his body (Figure 198).

Help can also be given with a knife, bayonet, or rifle butt (Figure 199).

Learning this kind of mutual help comes at the end of training and is built around various possible situations. In this training, the blows to be used should only be acted out. Full force is used only on punching bags or dummies.

The mutual help/rescue methods have to be performed at top speed, which minimizes the chance of harm to your comrade.

Naturally, there are instances where help indicated takes in throwing, twisting limbs, and even shooting. The simplest and most efficient rescue methods are surveyed and described in the following sections.

Throws

The throws we examine are not so different from the sport or martial arts kind and are means of defense, escape, and attack. In combination with defensive tactics, they serve as a means for sustaining single combat with an enemy. The theory, technical aspects, and methodology are found in many written sources here and abroad.

The most used and most effective throws for unexpected encounters and single combat are the backward trip/throw, the forward throw over the shoulder, and the throw, forward or backward, by jerking the legs out from under. They are simple to coordinate with other movements. Performing them proficiently is achieved through much repeated practice. These throws must have a place in all (physical) training. Practicing throws should be done in all kinds of daily dress, on streets, wearing flak jackets, in combination with other exercise, and after regular training.

A throw should be successful; that is, after it is done, an opponent should be lying on the ground. Success must be attained even at the expense of a beautiful, technical flawless

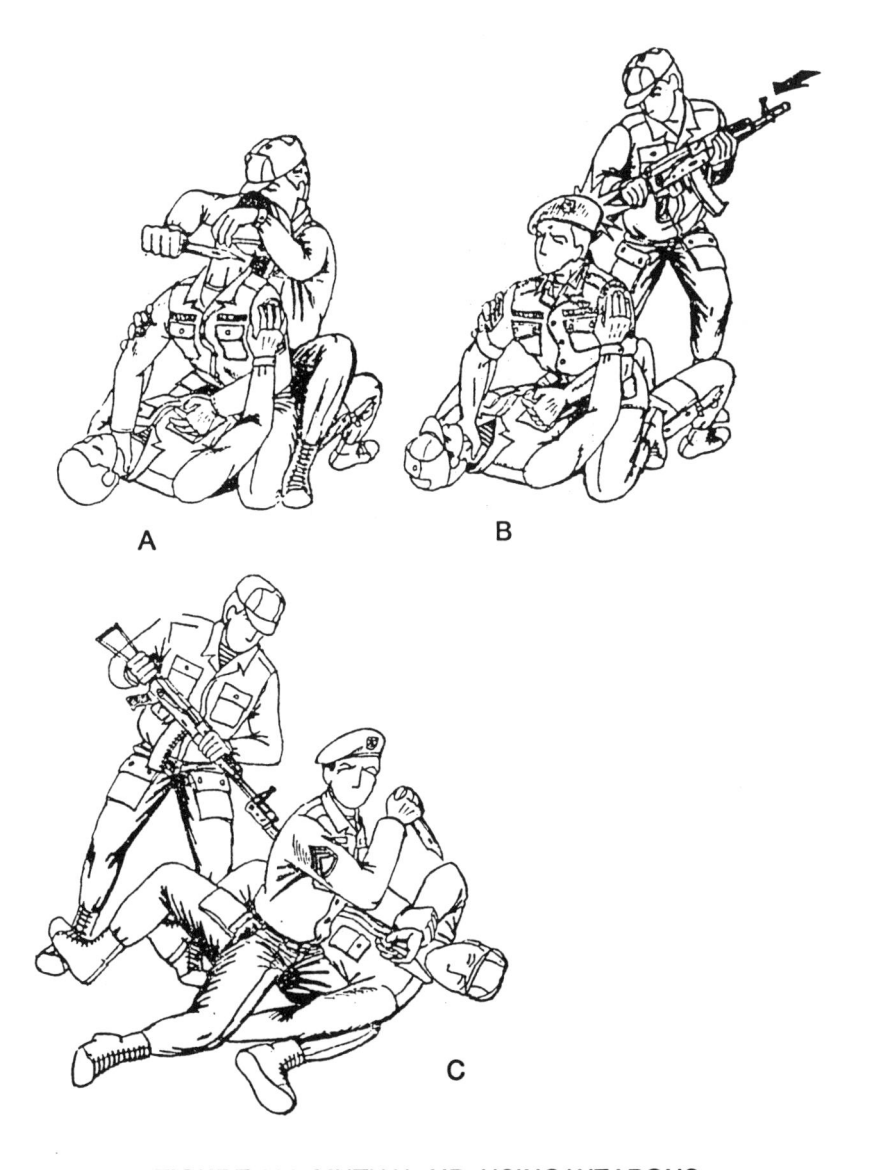

ESCAPING FROM AND FIGHTING OFF PHYSICAL ATTACKS; MUTUAL AID; THROWS

FIGURE 199. MUTUAL AID, USING WEAPONS
A—KNIFE
B—GUN BUTT
C—BAYONET

performance. It is important to accomplish throws confidently, even in such inconvenient situations as with a rifle slung across the back or chest. This means regular practice of three to five minutes duration in scheduled and individual training, morning and evening. Dangerous throws, in particular, have to be learned—and they are practiced only

FIGURE 200. FORWARD THROW OVER THE SHOULDER

FIGURE 201. DYNAMICS OF A FORWARD THROW

with dummies.

The forward throw over the shoulder (with one hand lifting the opponent's armpit) is performed with knees flexed, and with the enemy landing on his head (Figure 200).

In single combat, this throw can be done after a preliminary kick to the crotch or with a concluding kick, as shown in Figure 201.

This kind of throw is used against an attack from behind, holds on the neck, holds that pin the arms to the body, and so on. As stated earlier, in combat conditions an enemy is thrown without any regard for his safety: onto the ground, onto some physical object, onto his head. But in learning situations, all involved have to take care.

Throws that involve grabbing an opponent's legs (Figure 202a) are performed together with preceding blows that deter or stun, such as a head butt to the nose or a physical object to the head.

FIGURE 202. THROWING AN ENEMY
A—LOWER LEG GRAB
B—TRIP, WITH AN ELBOW TO THE FACE

A throw across the thigh, with an elbow to the face (Figure 202b) or fingers in the eyes or nostrils, serves as an efficient way to escape holds or as a defense against blows.

The throw across the thigh is based on the one used in martial arts or sport, but in the one used here, the right hand does not grab the opponent's clothing, but his throat. His head is first pushed up from his shoulders, and to the left and down, to throw him onto his head (Figure 203).

A tripping throw with an attack to the crotch or throat is also very effective in single combat (Figure 204).

After the enemy is down, kicks to the groin follow (Figure 205).

FIGURE 203. THROW ACROSS THE THIGH

If your fight is by a railing or a fence, it is very much in your interest to trip or throw your opponent so that his head or back hits such obstacles (Figure 206).

You can drop an enemy to the ground by hitting him in the back from behind or by jumping on him from above—for instance, when he is emerging from a dugout. Once he is down, tie him up or kill him with a cold weapon (Figure 207).

For single combat, and especially for practicing throws, it is very important to work with a training scheme. To work out a

A   B

FIGURE 204. TRIPPING FROM THE INSIDE OR OUTSIDE
A—ATTACKING THE THROAT
B—ATTACKING THE CROTCH

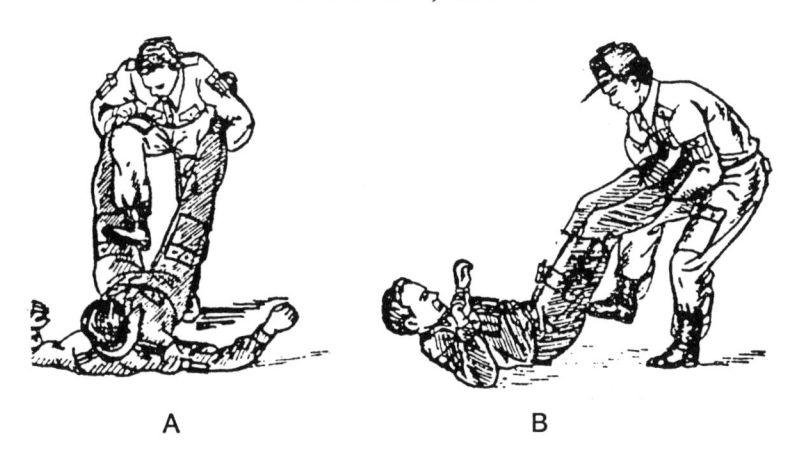

ESCAPING FROM AND FIGHTING OFF PHYSICAL ATTACKS; MUTUAL AID; THROWS

FIGURE 205. KICKS TO THE GROIN AFTER A THROW
A—FROM ABOVE
B—FROM BELOW

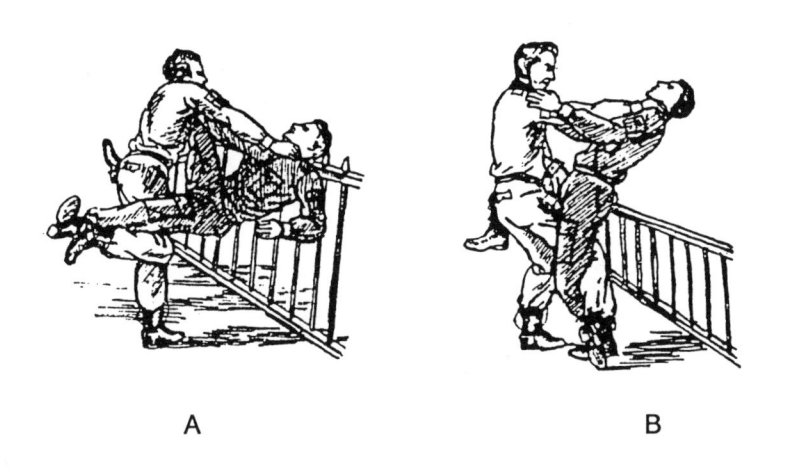

FIGURE 206. EXECUTING THE TECHNIQUE WITH AN ENEMY'S HEAD HITTING A FIXED OBJECT
A,B—TRIPPING BACKWARD

FIGURE 207. JUMPING AN ENEMY FROM ABOVE
A—JUMPING
B—PINNING

FIGURE 208. A SCHEME OF ATTACK AND DEFENSE EXERCISES

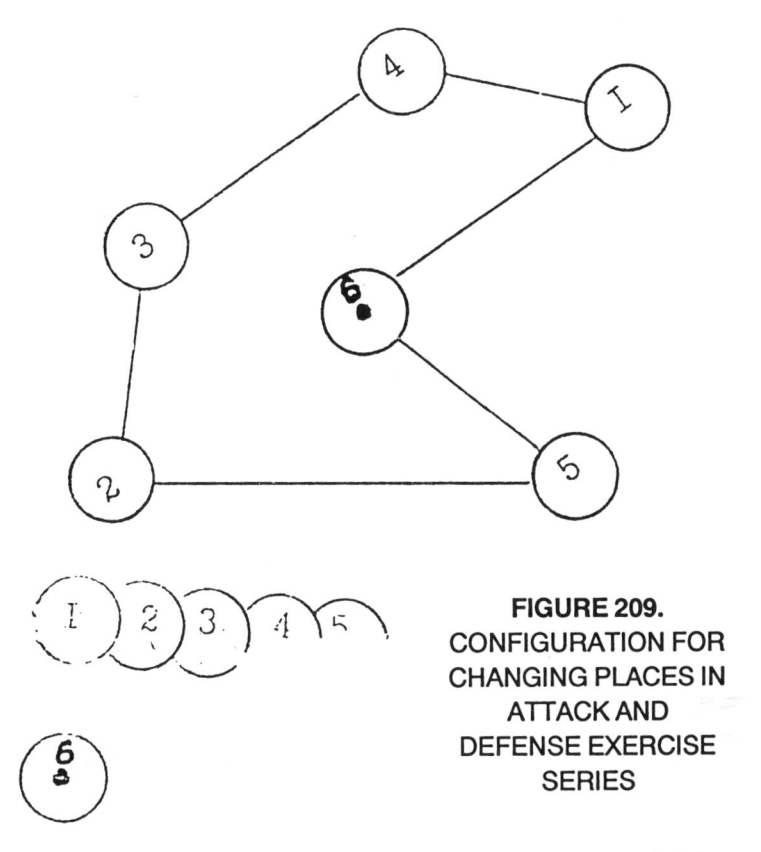

ESCAPING FROM AND FIGHTING OFF PHYSICAL ATTACKS; MUTUAL AID; THROWS

FIGURE 209. CONFIGURATION FOR CHANGING PLACES IN ATTACK AND DEFENSE EXERCISE SERIES

system of (self-) defense, there is a series of attack and defense exercises based on the position of the hands, blows, throws, and falls (Figure 208).

The series is worked through by a team of six individuals. One stands in the center of the circle and successively defends himself from his five teammates, who attack in turn. (All dangerous moves are only acted out.)

Here is how it goes:

1. Number 1 attacks number 6 (in the center), grabbing his shirt front, or delivers blows, and defends himself. Number 6 responds with a defensive throw or blow.

2. Number 2, exactly opposite from number 1, delivers a

kick. Number 6 defends with his arms, jerks number 2's foot, throws and then strikes with his foot or hand.

3. Number 3 acts out an overhand blow with a club or a knife. Number 6 defends with his left forearm, a blow, and a throw by tripping.

4. Numbers 4 and 5 move in at the same time, one trying for a chokehold from behind and the other going for the legs to make a throw. Number 6 beats off an attack from the front with his foot and throws the rearward choker over the shoulder, with a preliminary elbow to the stomach or chest.

After the whole series is completed, the one in the center moves to number 5's place, number 1 goes to the center, and the others move clockwise to their next position. The configuration for this practice series is for escaping from holds, defending against blows, and performing/avoiding throws, as shown in Figure 209.

In performing the series and changing places, everyone has to move carefully; after falling, each one gets back on his feet quickly to avoid injuries. One complete series should take one and one-half to two minutes. An excellent performance of exercise within a series is twelve seconds or less. The tempo of the number of blows is up to 220 blows per minute. For further development of self-defense, one of the directions of attack should not be indicated beforehand, but set up for unexpected action. The methodology of this training was tested and introduced into military training after 1976. In this regard, the best training is for the most physically developed and highly prepared personnel.

SELF-DEFENSE AGAINST AN ENEMY WITH A FIREARM

An attack by an enemy can be with a firearm (e.g., pistol, rifle, automatic weapon). Under the threat of being shot, a soldier can be searched, secured, and taken prisoner. The manual looks at different methods for application in situations where it is not possible to use a firearm for defense. The basic actions that can be taken are getting away, bowing to circumstances, or

ESCAPING FROM AND FIGHTING OFF PHYSICAL ATTACKS; MUTUAL AID; THROWS

fighting back with holds, blows, and throws. For the one defending himself against getting shot, there are objects that can be used, such as sticks, rocks, (even) tobacco, and so on. As a rule, an assault with a firearm consists of a sequence of set actions: threat, approach, waving or aiming the weapon, and immediate use, i.e., hitting or shooting.

Assessing a situation and choosing tactics and means of self-defense must generally be done in the stage before an actual attack. There can always be exceptions to the rule; all depends on the moment.

BASIC METHODS OF DEFENSE AGAINST A FIREARM AIMED FROM IN FRONT

You can defend yourself against an enemy with a pistol who is right in front of you—about one pace away—by kicking the gun out of his hand. After the kick (with the left foot from the

FIGURE 210. KICKING THE PISTOL FROM AN ENEMY'S HAND

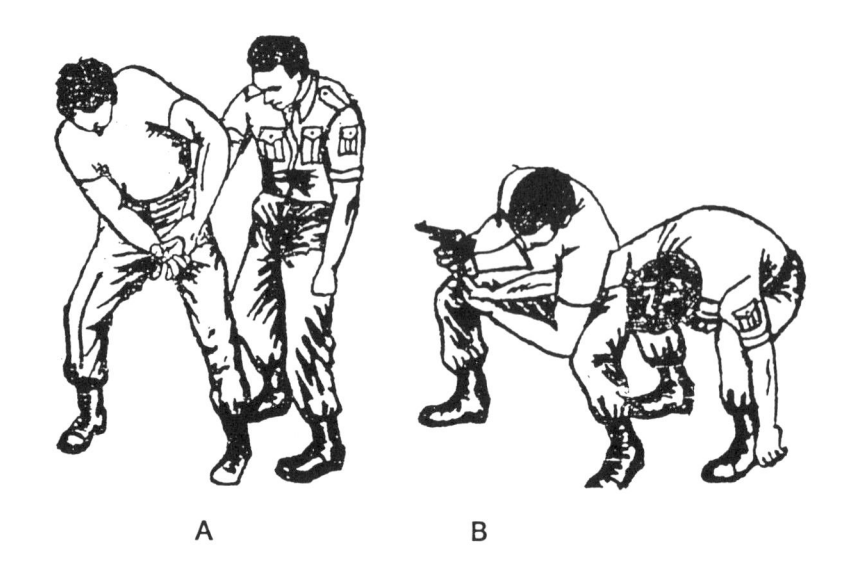

FIGURE 211. BENDING THE WEAPON
HAND/ARM TOWARD YOU
A—GRABBING THE HAND
B—BENDING

side or with either foot from below), move to close with him from the side for a hand-to-hand engagement. The kick is shown in Figure 210. To enhance the defense, use some physical object to hit him in the face.

If the distance to the weapon hand is less than one pace, then the best defense is to grab and twist the hand inward: step to the left and forward, turn to the right, go at your enemy from his right, and grab the weapon hand. Twist his hand toward you and disarm him. Twist the arm behind his back, inflict some damage, and tie him up and/or take him away.

This method is simplest and most effective when done immediately as the pistol is pointed at you from point-blank range. Right after "Hands up!" raise your hand edge to the left or right of the muzzle, simultaneously grabbing and pushing the weapon hand aside. Without letting go of that hand, use your free hand to knock away or take the pistol. You have to get it as

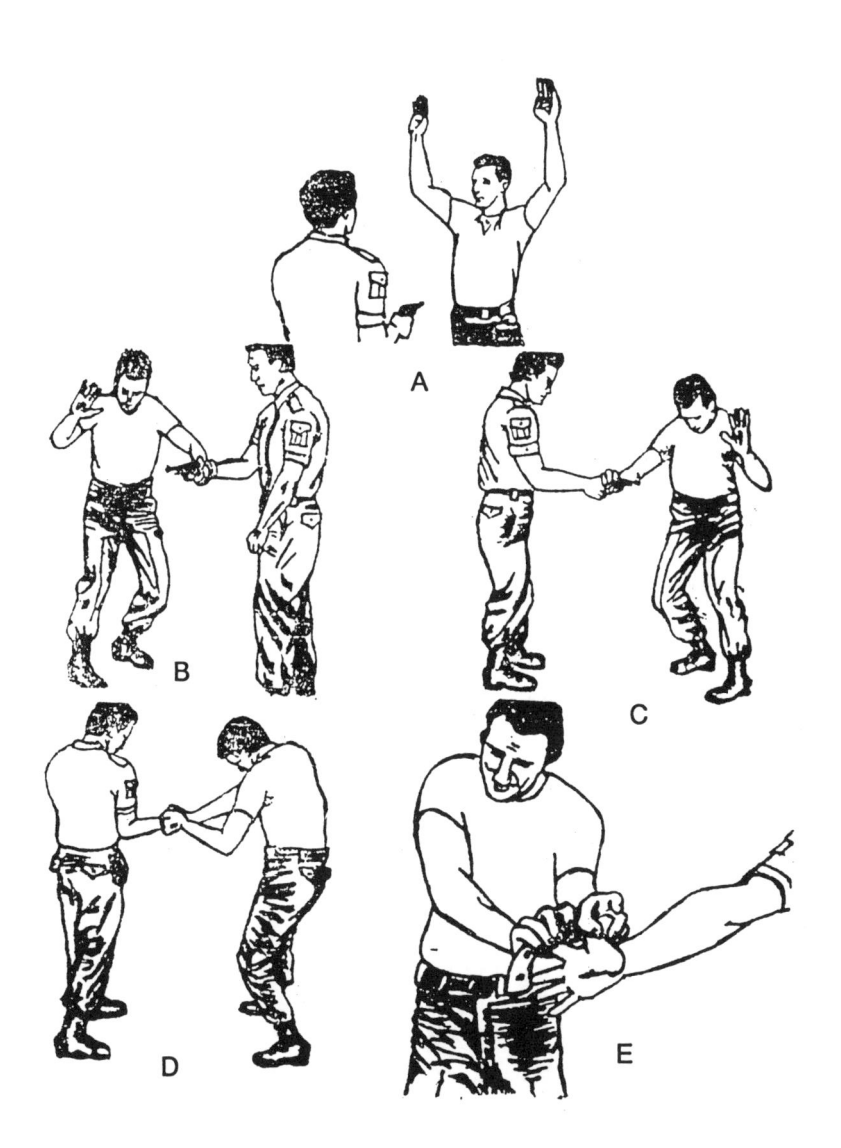

FIGURE 212. DEFENSE AGAINST AN ASSAULT WITH A PISTOL
A—ASSAULT
B,C—GRABBING THE WEAPON HAND
D,E—TWISTING THE WEAPON HAND

quickly as possible. Along with all this, a kick can be delivered. This method and the hand used depend on the realities of the situation. Figure 212 illustrates the method.

Dealing with an enemy armed with a long-barreled weapon is discussed below.

Leftward Movement

Put your left foot forward, move to the left, and, turning to the right, push at and grab the weapon with your right hand. Move your right foot toward the assailant's left foot and immobilize the weapon. After you have hold of the weapon, kick the assailant in the crotch with your left leg, hit him in the head with the butt of the weapon, and disarm him.

Rightward Movement

Advance your right foot, move to the right (next to the bayonet point or the muzzle) and, turning to the left, push at and grab the weapon with your left hand. Move your left foot toward the attacker's left foot. Deliver a blow with your right hand to his face or throat, and kick him in the bend of the knee with your right foot or knee. Then disarm him.

Figures 213 and 214 show the execution of these methods, depicting pushing aside and seizing the weapon, immobilizing it under the defender's arm, and a fist to the enemy's crotch or an elbow to his face.

Physical objects such as headgear, jackets, briefcases, and so forth can be used for defense. They can be thrown or swung into an enemy's face or against a pistol. Then you can close with him to use the methods just described. For teaching these methods and for psychological conditioning, the PM pistol with blanks or a starter's pistol can be employed. When kicks to a hand are practiced, the target should wear a boxing glove to guard against bruises and/or injuries.

BASIC METHODS OF DEFENSE AGAINST A FIREARM AIMED FROM BEHIND

To deal with a pistol pointed at you from behind, you can

ESCAPING FROM AND FIGHTING OFF PHYSICAL ATTACKS;
MUTUAL AID; THROWS

FIGURE 213. DEFENSE AGAINST ASSAULT WITH A LONG-BARRELED WEAPON
A—PUSHING, GRABBING THE WEAPON
B—IMMOBILIZING THE WEAPON
C—PUNCHING THE CROTCH

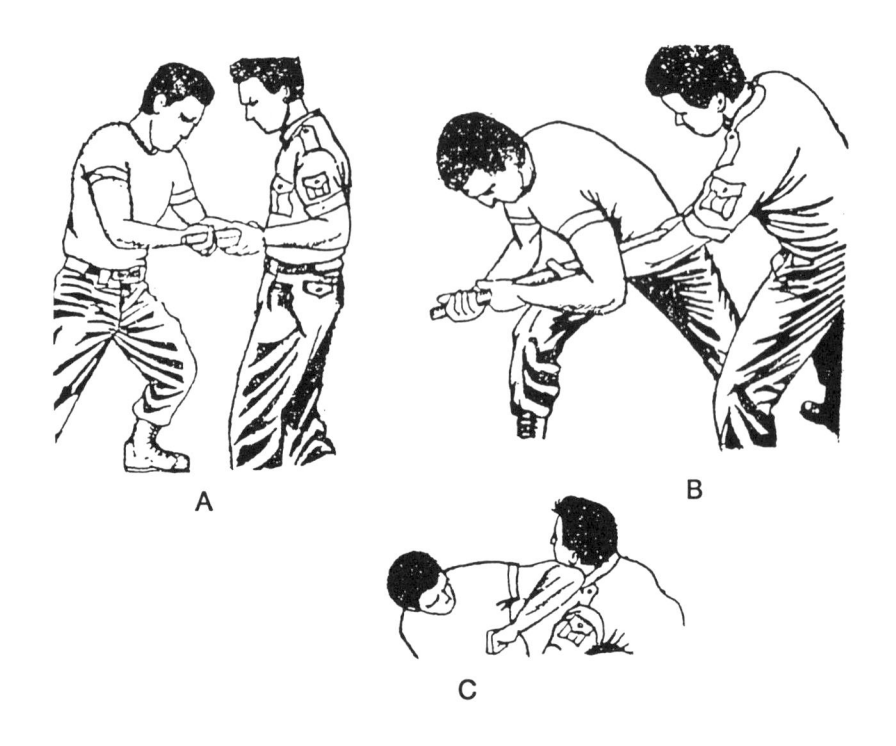

FIGURE 214. TAKING A LONG-BARRELED WEAPON AWAY
A—CLOSING IN
B—GETTING THE WEAPON UNDER THE ARM
C—ELBOW TO NECK OR THROAT

turn around to the right to seize and bend the weapon hand/arm outward or inward (Figure 215). You can turn around to the left, get the weapon arm under your left arm, and hit your assailant with your right hand or trip him backward.

Figure 217 shows the actions for defending against a long-barreled firearm aimed from behind. In this, you get the weapon under your left arm and deliver a blow with your right hand or arm or, grabbing your assailant's shirt front, trip him backward to the ground and kick him. Keep the weapon locked under your arm in all this.

Defending against and disarming an attacker are basically the

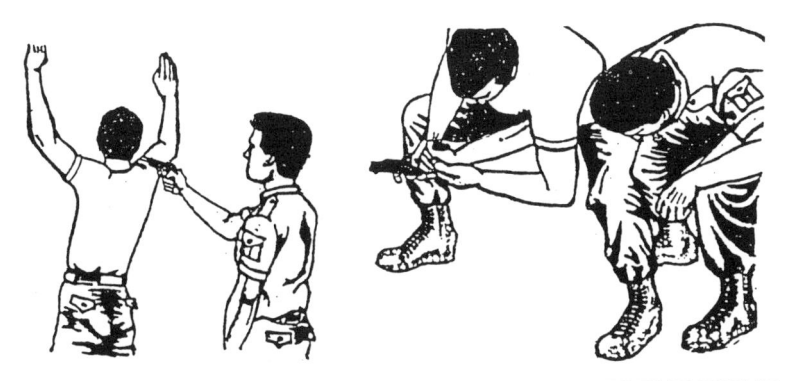

ESCAPING FROM AND FIGHTING OFF PHYSICAL ATTACKS; MUTUAL AID; THROWS

FIGURE 215. DEFENSE BY TURNING AROUND TO THE RIGHT

A

B

FIGURE 216. DEFENSE BY TURNING AROUND TO THE LEFT

C

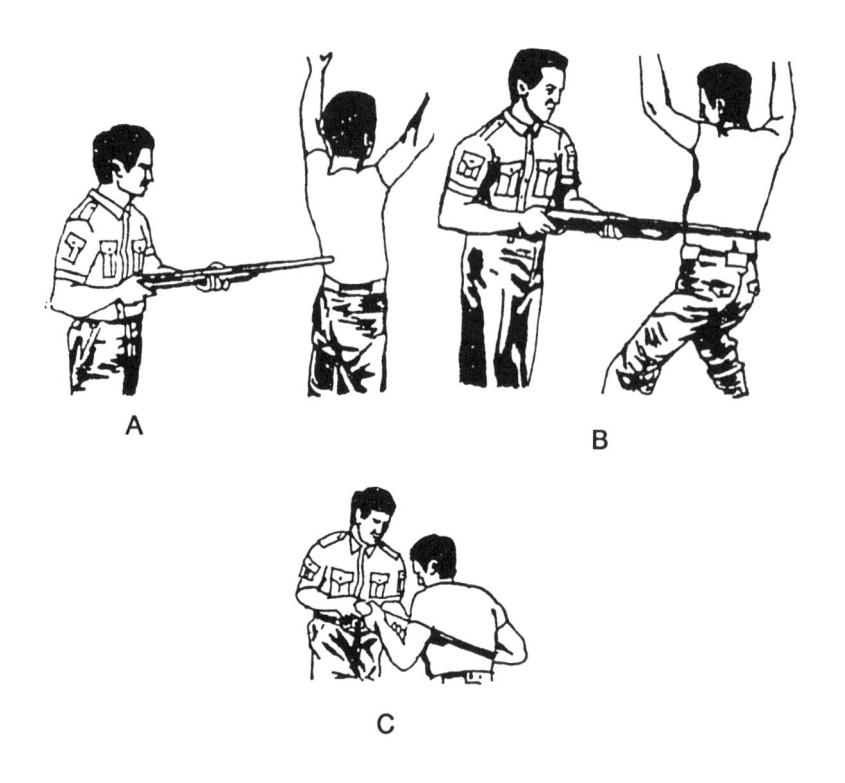

FIGURE 217. DEFENSE AGAINST A LONG-BARRELED
FIREARM BY TURNING AROUND TO THE LEFT
A—ASSAULT
B—TURNING AROUND
C—GETTING THE WEAPON UNDER THE
ARM AND PUNCHING THE CROTCH

same in assaults either from in front or from behind. Figures 218 and 219 show the techniques and dynamics.

In the defensive moves where the left foot is moved leftward and the left hand is used to push the long-barreled weapon aside, the right hand is used to grab the butt of the weapon, to wrest it away.

DEFENSE AGAINST COLD WEAPONS

Cold weapons include, among others, knives, axes, awls, and spikes. We will examine defense against one of this array—the knife.

In single combat, of primary importance is how the knife is held by an attacker. If the point is up (away from the thumb), then a stroke can be expected from below or from the side, at the stomach, side, or neck. If the point is down (away from the little finger), the blow will be overhand, backhand, or a combination of feinting with a stab to the chest or midriff (Figure 220).

Such blows can be expected especially when an attacker is coming from the side. Overhand blows have to be warded off carefully because they are hard to defend against except by backing away, using kicks or physical objects already in hand at the moment of attack.

Defense techniques depend on many factors and must be as varied as possible.

The classic defense technique of twisting the assailant's weapon hand and arm are effective only when employing a powerful kick, or striking with the fingers or head at a vulnerable point of his body.

The use of disabling tactics in a situation of an overhand stab with feinting, e.g., pain-producing arm twisting, depends on several things: the attacker's size, the knife's position, the type of weapon (double- or single-edged), personal equipment, and agility, strength, and bravery as well. Psychological toughness plays a vital part in fighting an enemy armed with a real weapon. Thus, the education of armed forces personnel has to put aside the usual, everyday methods of training: static teaching and lecturing, artificial situations, and dummy weapons. Practical applications have shown that carrying out exercises on the street at any time of year, wearing regular clothes and equipment, and using real cold weapons disciplines the trainees, establishes a connection between personal defense and tactics, heightens interest, and toughens the spirit. But the instructor who thinks carefully about this course of training must retain

FIGURE 218. WRESTING A WEAPON
FROM AN ENEMY'S HAND

overall control, looking after each student.

Inasmuch as the basic techniques of defense are described in detail in technical literature, this manual will dwell briefly on overhand stabs and those from the side and below. Of interest, at this level, is a discussion of the utilization of physical objects for self-defense.

The following methods can be employed in defending

FIGURE 219. TWISTING A WEAPON FROM AN ENEMY'S GRASP

against a knife attack: blocking the assailant's forearm, grabbing the wrist of his knife hand, kicking his groin with the foot or knee, bending his arm or hand, and putting a lock on his knife hand/arm (overhand, underhand, and so on).

If an attacker is not much taller than the defender and is not armed with a double-edged knife, but has instead a stiletto, a *kindzhal* [a long dagger of Caucasus Mountains tribes], an awl, or suchlike, a hand/arm lock is an effective means of defense. This is illustrated in Figures 221 and 222. What has to be done in the final stage of fighting an enemy armed with only a pointed or single-edged weapon is to injure him. An assailant battered by a well-aimed blow to the neck, face, side, stomach, or back will be incapacitated by shock.

THE OVERHAND ARM/HAND LOCK ("OVERHAND ARM KNOT")

With your right hand (thumb side down), grab the attacker's right (weapon) hand by the wrist in an overhand grip, or block

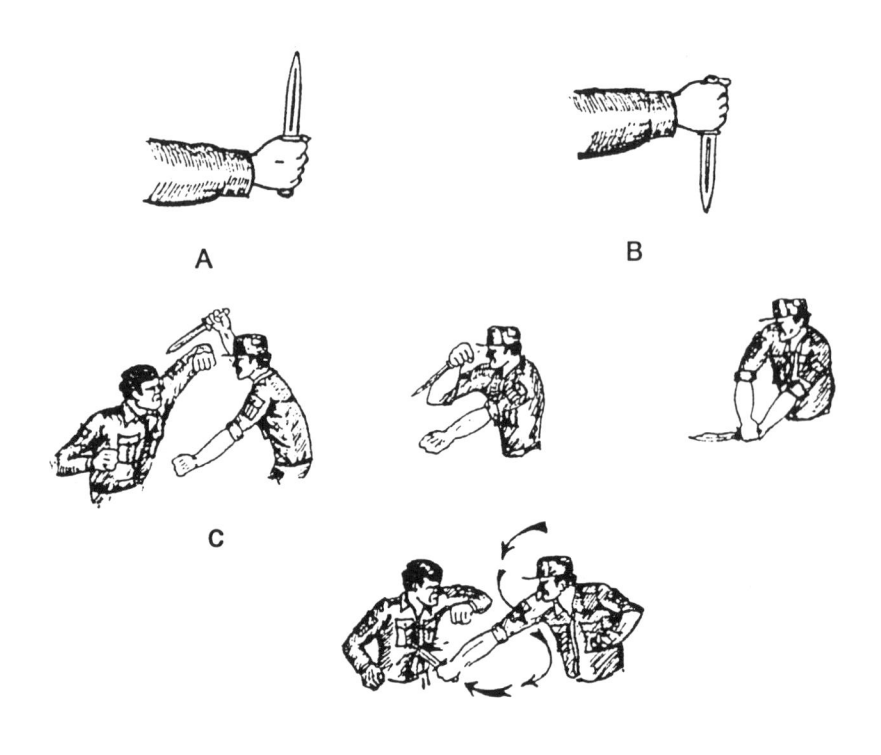

FIGURE 220. KNIFE POSITIONS
AND STABBING TECHNIQUES
A—POINT UP
B—POINT DOWN
C—DYNAMICS OF AN OVERHAND STAB
TO THE STOMACH WITH FEINTING

his right forearm and seize the clenched right hand. Pulling it toward you, put your left arm behind his right arm at the bent elbow. Raise his arm and grasp your own right wrist with your left hand ("tying the knot"). Turn leftward, kneeing your assailant in the groin and knocking him down.

If necessary, use your right hand on his weapon hand, grab his sleeve at the elbow, and turn him face down to tie him up and move him out. The techniques are shown in Figures 221 and 222.

ESCAPING FROM AND FIGHTING OFF PHYSICAL ATTACKS; MUTUAL AID; THROWS

FIGURE 221. USING THE "OVERHAND KNOT" AGAINST A KNIFE ATTACK
A—GRIPPING THE KNIFE HAND
B—CONCLUDING THE ACTION

INWARD ARM TWIST AGAINST AN OVERHAND STAB

Move your left foot forward; meet the descending knife arm of your assailant with your left forearm. Gripping his wrist or forearm with your right hand, perform the inward arm twist. That is, with a turn to the right, stretch his knife hand/arm out as if to bring it toward your right ear, grip his arm under your upper left arm, and align your feet with your assailant's. Keeping his arm parallel to the ground, apply pain and disarm him with your free hand. Twist the arm behind his back, secure him, and move him out. Further, once you have gotten ahold of his knife hand (at the wrist), you can use an inward twist of his hand and arm (Figure 223). To ensure complete success, a kick in his groin is essential.

The technique of using the forearm to block a blow is described in detail in various technical manuals. It is not very different from the techniques illustrated here. If an attack is made with a pointed weapon, then, after stopping the stroke with your left forearm and grabbing your assailant under the armpit of his weapon arm, he can be thrown by your turning around to

FIGURE 222. USING THE "UNDERHAND KNOT"
AGAINST A KNIFE ATTACK
A—THE ATTACK
B—GRIPPING THE HAND/ARM
C—BACKWARD TRIPPING TO CONCLUDE THE ACTION

the left and following through with a throw—or he can be put down on the ground with a trip.

UNDERHAND STABS

A forearm or two crossed forearm (right over left during training) should be used against an underhand stab to the stomach or groin. Subsequent action depends on the situation:

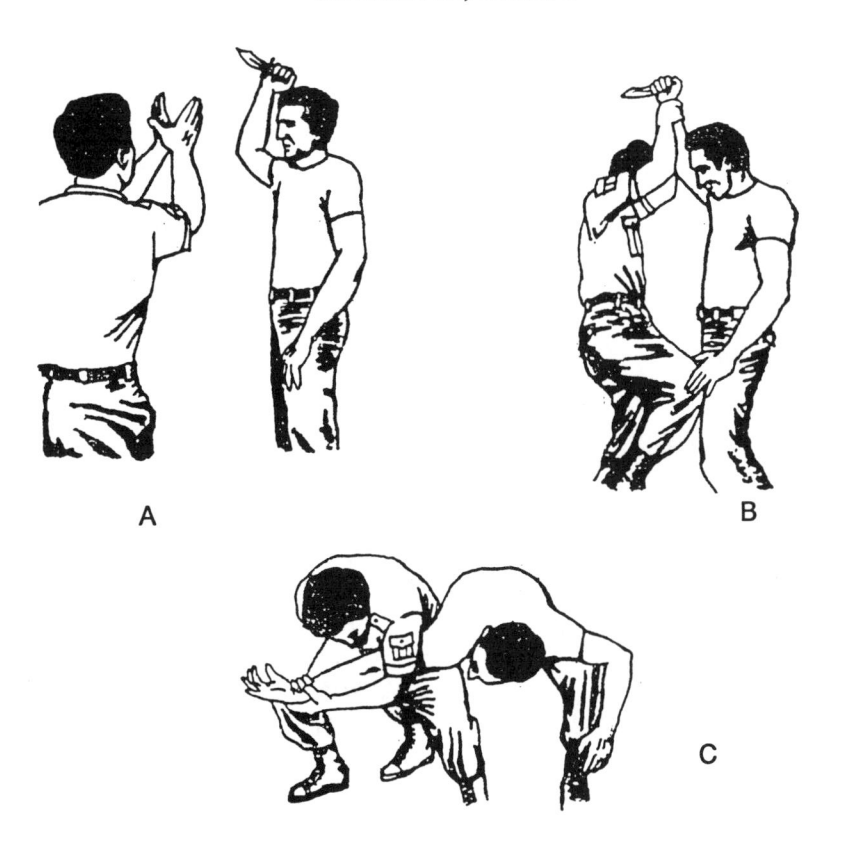

ESCAPING FROM AND FIGHTING OFF PHYSICAL ATTACKS;
MUTUAL AID; THROWS

FIGURE 223. DEFENSE AGAINST AN OVERHAND STAB
A—GETTING AHOLD OF THE WRIST
B—KICKING THE GROIN
C—INWARD TWISTING THE HAND OR ARM

either a kick to the assailant's crotch or twisting his weapon arm behind his back. For the latter, advance your left foot to meet the attacker with your left arm and bend forward slightly at the waist. Grab his weapon arm sleeve at the elbow with your right hand and knee him in the crotch. Folding him toward you, twist his arm behind his back, disarm him, hurt him, secure him, and move him out.

Figure 224 shows a variation in which you stop his weapon

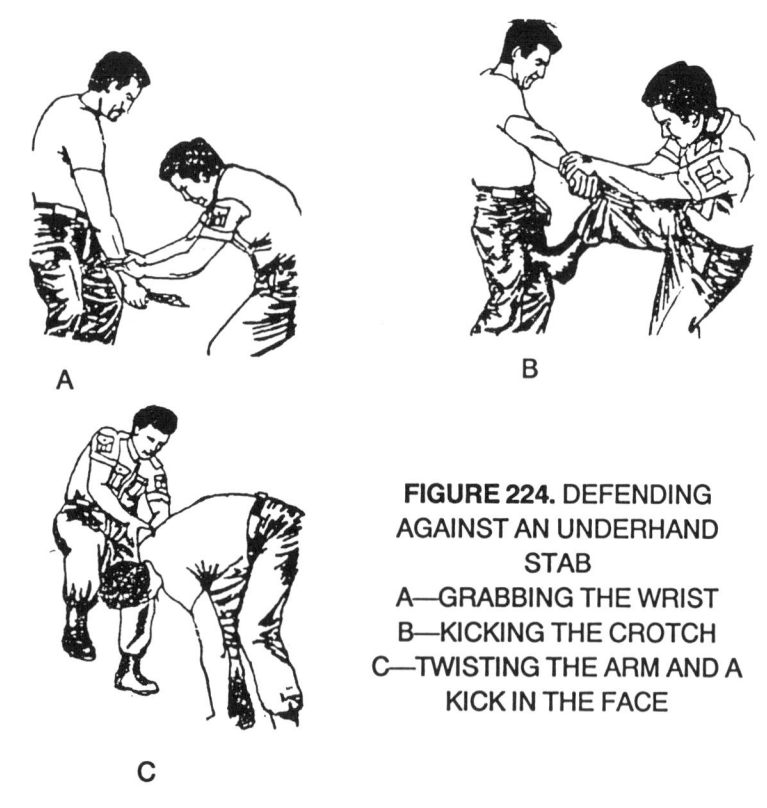

FIGURE 224. DEFENDING AGAINST AN UNDERHAND STAB
A—GRABBING THE WRIST
B—KICKING THE CROTCH
C—TWISTING THE ARM AND A KICK IN THE FACE

hand by grasping it.

Continuing, deliver a blow to his crotch, which bends him toward you. From the position taken in 224c, twist the weapon hand and carry out an inward arm twist, or disarm him along with a kick in his face.

BACKHAND AND LATERAL STABS

Use your forearms against backhand or lateral stabs, and firmly grasp the assailant's weapon hand (Figure 225). Also, use diversionary kicks to his groin, and employ pain-producing techniques such as an inward arm twist, as shown in Figure 226. In the case of a lateral stab, block it with both forearms and use the "overhand knot," the inward twist, or a kick to the attacker's

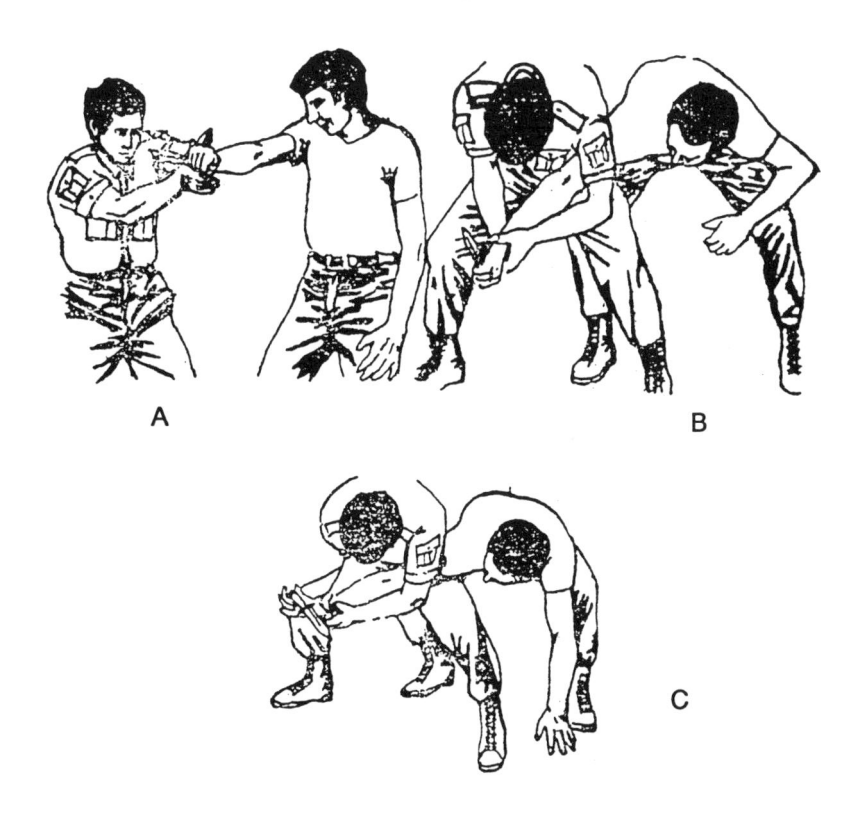

ESCAPING FROM AND FIGHTING OFF PHYSICAL ATTACKS; MUTUAL AID; THROWS

FIGURE 225. DEFENDING AGAINST A (BACKHAND) LATERAL STAB
A—GRIPPING THE WRIST
B,C—INWARD ARM TWIST

crotch with your right knee.

Figure 226 shows a lateral stab where the knife point is away from the attacker's little finger.

Our own literature on methods in such attacks recommends a defense with the left foot moved forward and a turn to the right with the forearm out (if there is a lateral stab from the right, then the left forearm). The defender does not seize the knife hand—

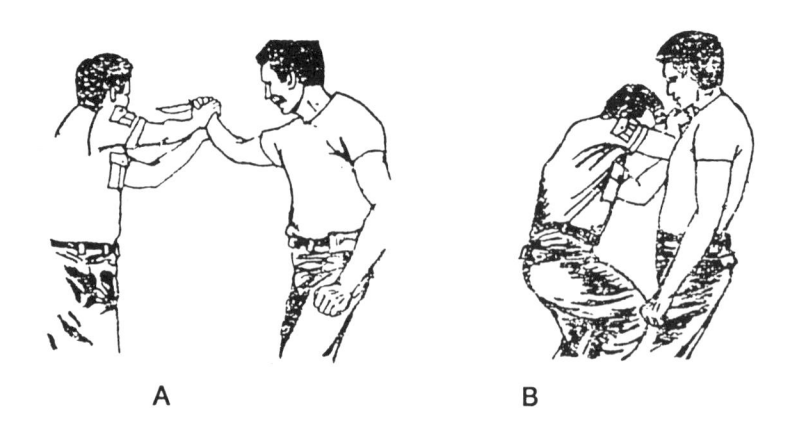

FIGURE 226. DEFENDING AGAINST A LATERAL STAB
A—GRIPPING THE WRIST
B—INWARD OR OUTWARD ARM TWIST

but this is wrong, dangerous, and not how it is done "at the front." During training in the gym and with practice or in the first stage of teaching self-defense, that method may be taught, but it is no good in the real world at all! The weapon hand has to be seized immediately, and best of all, right when it is moving. Then, the action to take is determined by circumstances.

A knife blow is particularly dangerous, whether a puncture or a cut. It generally strikes the chest or stomach, producing a puncture wound. Defense against such includes backing away, preemptive fingers in the assailant's eyes, kicks in his groin, and physical objects to his head or hand. A knife stroke is brief.

Move your left foot forward and leftward so as to meet the attacker's knife hand/arm with forearm blocks and by grasping with your hands. Divert the blow downward and to the right. Then get hold of the knife hand with your right hand. Holding on to him with your left hand, kick him in the groin and twist his arm outward. Disarm him or kick him; twist his arm(s) and tie him up, or march him off with his arm(s) twisted behind his back.

Averting the attack is taken care of by kicking the held knife, especially if the shoe or boot has a heavy, thick sole.

ESCAPING FROM AND FIGHTING OFF PHYSICAL ATTACKS;
MUTUAL AID; THROWS

A   B

FIGURE 227. DEFENDING AGAINST A STAB OR SLASH
A—BACKING OFF
B—KICKING THE KNIFE HAND

During a knife attack, which can be aimed at a hand, knee, or groin, reaction is determined by the immediate situation: you can close with your assailant for a fight to disarm him, or you can jump back t take action for a sure self-defense (Figure 227).

If such is available, it is a good idea to throw something into an attacker's eyes or face.

You can knock the knife away, to the left, right, or down, with the barrel of your automatic rifle. After that, point the muzzle at him, jab him with the barrel, or hit him with the butt or magazine. If you find yourself right next to an enemy who pulls a weapon out of his pocket or from under his arm, take the following action. Get right up to him and immediately force his weapon arm/hand against his body with your left hand or both hands. Then use your right hand as is appropriate to the situation: that is, deliver a blow or grab his weapon arm sleeve a little above the elbow to twist his arm, or grab his weapon hand and pull him toward you so that you can twist his hand outward. In any event, once his arm is solidly held, give him a kick in the crotch.

SELF-DEFENSE USING ADDITIONAL MEANS

The best means of defense against an assailant with a firearm or cold weapon is anything available. The effectiveness of this kind of defense lies in the unexpected use of such means—in combination with diverting actions—which facilitates going on to the better-known methods of self-defense: evasive moves, backing off, blows, twists, or throws. Blows, of course, are done with the head, fists, fingers, feet and legs, and so forth. These blows are used on an enemy to prepare the way for follow-up, more complex techniques. In addition to blows, throws, and the rest, in combat situations the teeth are sometimes used too.

BASIC TECHNIQUES

Just when an attacker is swinging a club down at you, block the blow with your forearm and poke the fingers of the other hand in his eyes. Then exploit the situation: carry out a kick to the groin or a throw (Figure 228).

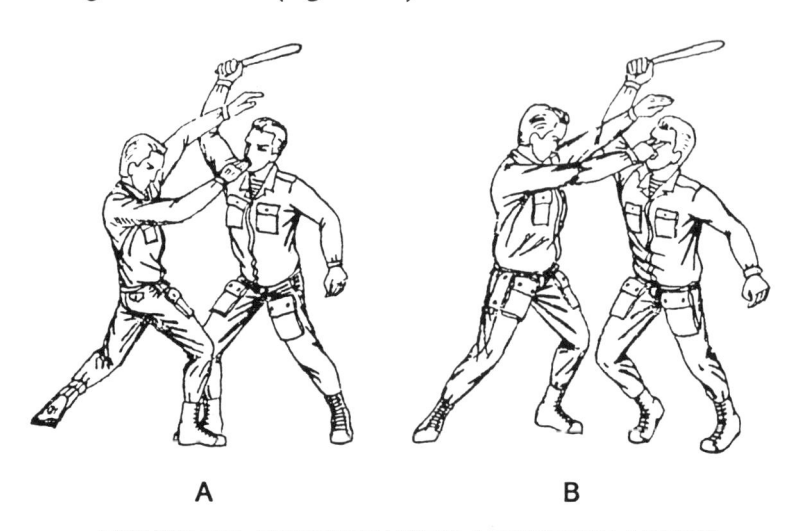

FIGURE 228. DEFENSE USING ADDITIONAL MEANS
  A—A HAND TO THE NOSE
  B—FINGERS IN THE EYES

ESCAPING FROM AND FIGHTING OFF PHYSICAL ATTACKS; MUTUAL AID; THROWS

FIGURE 229. DEFENSE WITH A HEAD BUTT TO THE FACE

FIGURE 230. DEFENSE WITH A FOREARM AND A STEP FORWARD

After using your forearm and taking a step forward, grab the attacker's shirtfront with your right hand, deliver a sharp blow with your head to his face or nose (Figure 229). After this, his head drops forward. This is the best moment to give a decisive kick to his crotch.

Use your middle and index fingers in his eyes; this is taught in independent training in supplementary techniques.

As a variant to defense against overhand or lateral blows with a club or other weapon, Figure 230 illustrates answering blows to an assailant's chest and face (holding a rock, piece of iron, or a chunk of wood in your hand). Anything lying around that can be picked up may be used to counter such attacks.

To defend against an enemy armed with any kind of weapon, it is especially effective to use anything that weakens his attack, e.g., sand, dirt, tobacco, and other dry substances. Get back from your assailant, get a handful of sand or suchlike, and step forward to throw it in his face (Figure 231).

The left hand can be used in feinting at your opponent's face, using the fingers or a fist, to distract him and turn the situation to your advantage. Then the enemy's weapon can be turned aside or a kick to his crotch performed.

Headgear can serve as a convenient item of defense against a cold weapon. This includes billed caps, hats, berets, garrison caps, and so on. The back of any head wear can be reinforced with a metal disk (lead is best) 30 x 20 x 1.5 millimeters in its dimensions, which fits just right. Snatch your cap by its bill or visor off your head and hit downward or from the side against a weapon hand (Figure 232).

Briefcases, valises, and satchels can also be used for defense or attack. Methods are shown in Figure 233.

FIGURE 231. THROWING AVAILABLE STUFF (SAND) IN THE FACE

A   B

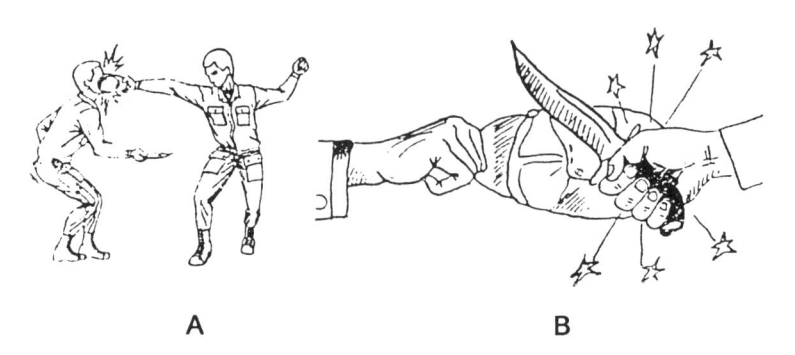

FIGURE 232. USING HEADGEAR FOR DEFENSE
A—HEAD BLOW
B—A BLOW TO THE HAND

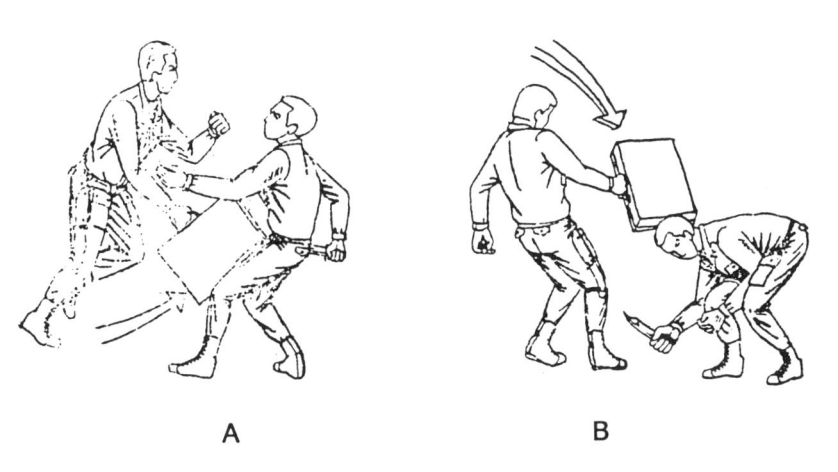

ESCAPING FROM AND FIGHTING OFF PHYSICAL ATTACKS; MUTUAL AID; THROWS

FIGURE 233. LUGGAGE AS A MEANS OF DEFENSE
A—BLOW TO THE CROTCH
B—BLOW TO THE HEAD

Methods of Defense

When an enemy threatens with a cold weapon, use a quick, accurate response from below, swinging a briefcase or valise to his groin. To make him writhe with pain, bring it down on his head, following up with appropriate actions: closing with him, disarming, knocking him down, and so on. Another good means of defense is a belt with its buckle.

A folded umbrella works very well against an attack with an edged weapon. When encountering the threat of an attack, jab the umbrella in the assailant's face and close with him quickly to throw or disarm him (Figure 234), using kicks or twists.

A sports jacket or short, padded coat (*kurtka*) with keys and wallet in a pocket is also effective for self-defense (Figure 235). In an armed attack, the jacket can be swung into an assailant's face or against the weapon hand, and then other self-defense measures can be taken.

It is best to use a short, sharp swing from the side or overhand to the head and then to close with the attacker and act accordingly.

If, for example, keys, wallet, and other items are put in an

FIGURE 234. AN UMBRELLA AS A MEANS OF DEFENSE

outside pocket, a blow with the jacket is powerful enough to knock over a combat practice dummy or a chair with a heavy weight on it, or break laths and boards. For use in this way, it is best that the jacket be carried in one hand or slung over the shoulder.

The use of an entrenching tool as a means of defense, is written about extensively in military literature. Blows with it are delivered overhand, from the side, as slashes to the face, and so on. One way is shown in Figure 236.

In close combat, a firearm is an excellent physical object to use. Any part of the weapon can be utilized against a vulnerable spot on the body. As Figure 237 shows, the best target area is the face. After this blow, a kick to the groin would not be excessive.

Headgear can be used as a good means of defense against an

enemy with a firearm, especially if the metal disk is sewn in (as described earlier).

It has to be used immediately after "hands up!" is heard, just when your hands come level with the bill or brim. The headgear item is used as show in Figure 238. After this, move in on your enemy and take appropriate action.

Figure 239 shows how a cap can be used by throwing it from a few steps away. After the "hands-up" command, raising the hands slowly, suddenly snatch the cap off your head by its bill, and with a step forward, throw it in the enemy's face. Then kick his weapon hand, knocking the gun out of it. Take further necessary action: springing forward and throwing him to the ground, tying him up, and so forth are recommended.

FIGURE 235. A JACKET AS A MEANS OF DEFENSE

A length of rope divided into three strands and wrapped—of the nunchuk type—can be used for self-defense. To make a nunchuk-style rope, lay out the rope in three strands, using half their diameters or less. Wrap one strand tightly around the other two, perpendicular to the two. Alpinists use this wrapping method on the loose ends of their safety lines.

If a nail or other thin, hard object is woven into this rope, a blow with it to the head can be deadly. The unwrapped end of this rope may be about twenty centimeters long or a bit longer (Figure 240). Carry this prepared device conveniently in an inside pocket.

FIGURE 236. AN ENTRENCHING TOOL AS A MEANS OF DEFENSE

Effective use of physical objects in self-defense is ensured by practicing with them in the course of training. A practice dummy can be used as a target enemy. The dummy should be supported from behind by a partner, and the partner should wear a padded glove on the supporting hand to avoid serious injuries or bruises. Blows with physical objects should be acted out and not aimed near the partner's head, but at the body of the dummy, and should have full force behind them. Practicing these self-defense methods can be done during morning exercising or workouts—and should only be acted out.

For quick morale building and psychological-physical conditioning, the personnel of a team should engage extensively

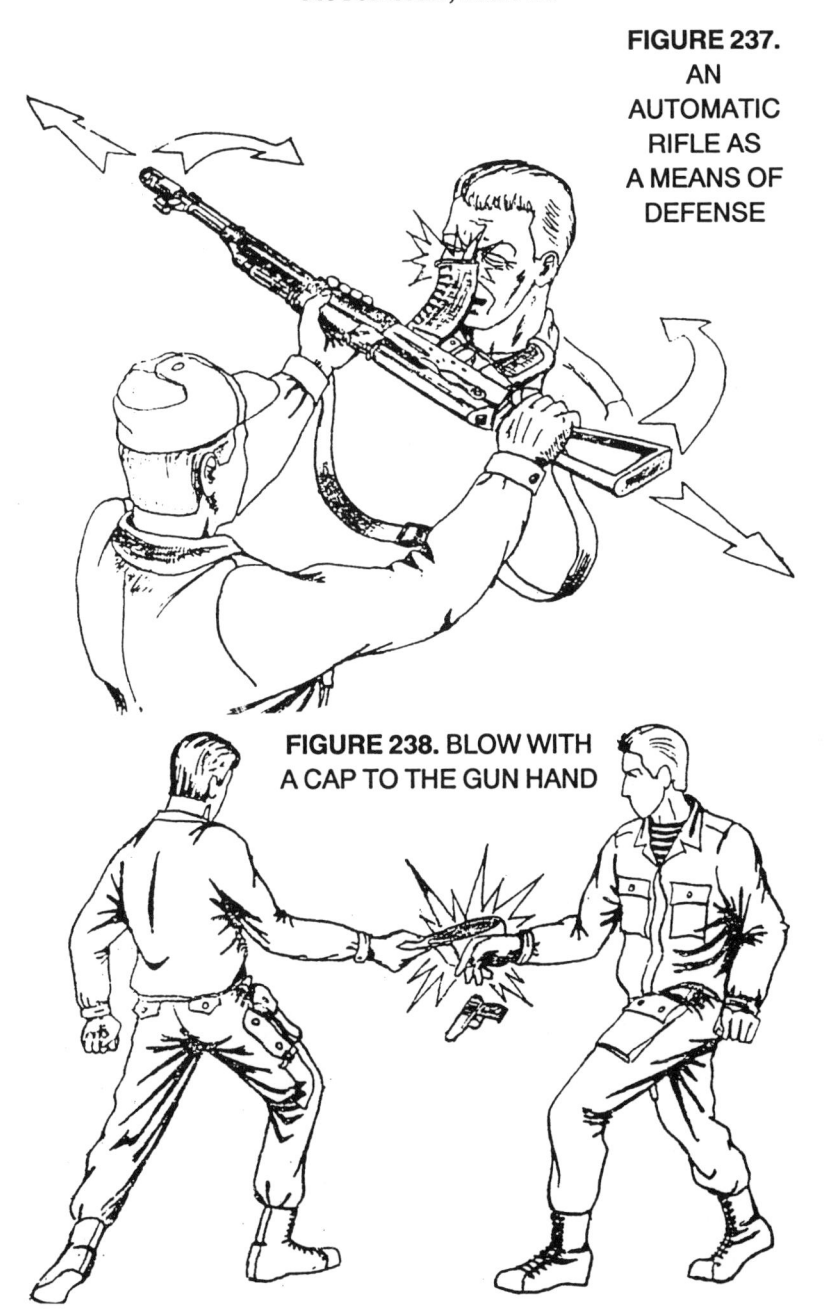

ESCAPING FROM AND FIGHTING OFF PHYSICAL ATTACKS;
MUTUAL AID; THROWS

FIGURE 237. AN AUTOMATIC RIFLE AS A MEANS OF DEFENSE

FIGURE 238. BLOW WITH A CAP TO THE GUN HAND

KGB ALPHA TEAM TRAINING MANUAL

FIGURE 239. THROWING A CAP AT THE HEAD AND KICKING THE WEAPON HAND

FIGURE 240. A ROPE AS A MEANS OF DEFENSE

FIGURE 241. SIMPLE PROTECTIVE DEVICES
A—BOXING HELMET
B—GLOVES
C—VEST
D—SHIN GUARDS
E—ARM GUARDS
F—PADDED SHIELDS

in Thai boxing matches and karate—two sixty- to ninety-second rounds of each are good. The blows and blocking them are to be learned and incorporated into training exercises. To avoid injuries, simple protective devices such as boxing helmets and gloves, soccer shin guards, padded shields, cups, and sleeveless padded vests should be obtained and adopted (Figure 241).

The blows shown in Figure 242 are dangerous, even in sports matches, and should not be attempted because of their complexity. Aside from that, someone trying fast, high kick to the head while fully dressed and without warm-up exercises will have his foot caught and will be thrown to the ground without being able to get up. But training carefully with a dummy, a punching bag, and other devices is best, because that kind of

KGB ALPHA TEAM TRAINING MANUAL

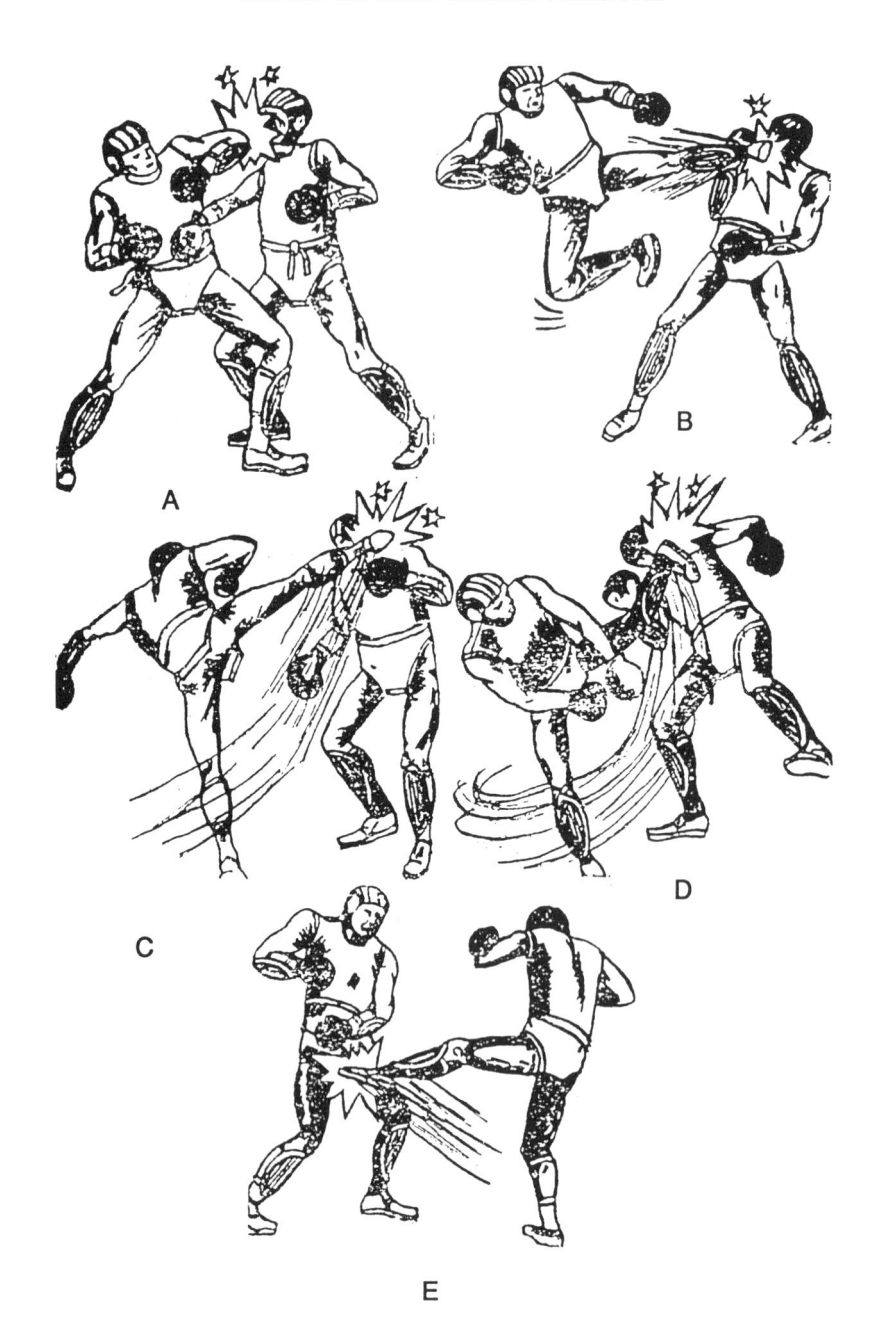

A

B

C

D

E

FIGURE 242 (LEFT). BLOWS USED ONLY ON DUMMIES, PUNCHING BAGS, AND OTHER DEVICES
A—ELBOW TO THE FACE
B—FOOT TO THE FACE
C—SIDE KICK TO THE HEAD
D—REARWARD KICK TO THE HEAD
E—KICK TO THE GROIN

activity develops speed and strength and produces balance and coordination.

Sparring matches refine the ability to estimate distance, hit a moving target actively defending itself, and expand the crucial qualities of morale (e.g., willpower, hardiness, decisiveness, and courage) so necessary for a serviceman.

Crotch kicks and elbow blows to the head and face are not allowed in sparring matches. These blows are not of the instructional-training kind; they are exclusively for combat.

Defense against an armed enemy should be learned from the additional methods and techniques of defeating an enemy and putting him in a state of shock, and used extensively.

ADDITIONAL WAYS TO DEFEAT AN ENEMY WITHOUT USING WEAPONS

Blows with the fingers and the edge of the hand to the eyes, neck, and ears, from behind or the side, and to the groin can disorient an assailant, force him to quit resistance, and sap his will.

Disorienting blows are those to the nose from in front or behind, and to the back, stomach, neck, and groin, as shown in Figure 244.

Defense against attacks encompasses kicks to the shin, kicks or stomps to the knee from the side or to the neck vertebrae (Figure 245). The most effective and hardest to oppose is the kick to the groin from below.

After using throws or arm twists, surveyed earlier and shown again in Figures 246 through 249 (these actions are not for sport but for real fighting, considering their ability to seriously injure),

FIGURE 243. ADDITIONAL METHODS AND TECHNIQUES
A—FINGERS TO THE EYES AND THROAT
B—BLOWS TO THE EARS
C—OVERHAND BLOW TO THE NECK
D—BLOW TO THE GROIN
E—BLOW TO THE THROAT WITH EDGE OF THE HAND
F—LATERAL BLOW TO THE NECK WITH
THE EDGE OF THE HAND
G—BLOW TO THE NAPE WITH THE EDGE OF THE HAND

ESCAPING FROM AND FIGHTING OFF PHYSICAL ATTACKS;
MUTUAL AID; THROWS

FIGURE 244. DISORIENTING BLOWS

a finishing-up blow should be used on an enemy to put him in a state of shock.

A personal firearm or cold weapon defeats an enemy right away, but kicks can "finish" an enemy too. The latter are not complicated and are shown in Figures 250 through 252.

Kicks are performed with not only the toe, but the heel, too, as shown in Figure 252.

In addition, kicking an enemy in the crotch when he is on his back is extremely effective. For example, after you have thrown an attacker by jerking his legs from the front, spread his legs as far apart as possible and then stomp on his groin. Also effective sometimes are kicks to the temple, jaw, and back of the head.

If a downed enemy moves his knees in the way of your attempted stomp to his groin, then shift your kick lower to strike the then exposed area. Simultaneously grabbing his arms, clothes, or hair accompanies blows to vulnerable points (Figure 253).

By pulling an enemy toward you, using some part of his body that you have hold on, you can bring him into contact with more powerful blows from your fist, forehead, or knee.

One example of a deadly method of defeating an opponent is a hold that damages his neck vertebrae, rupturing the cartilage.

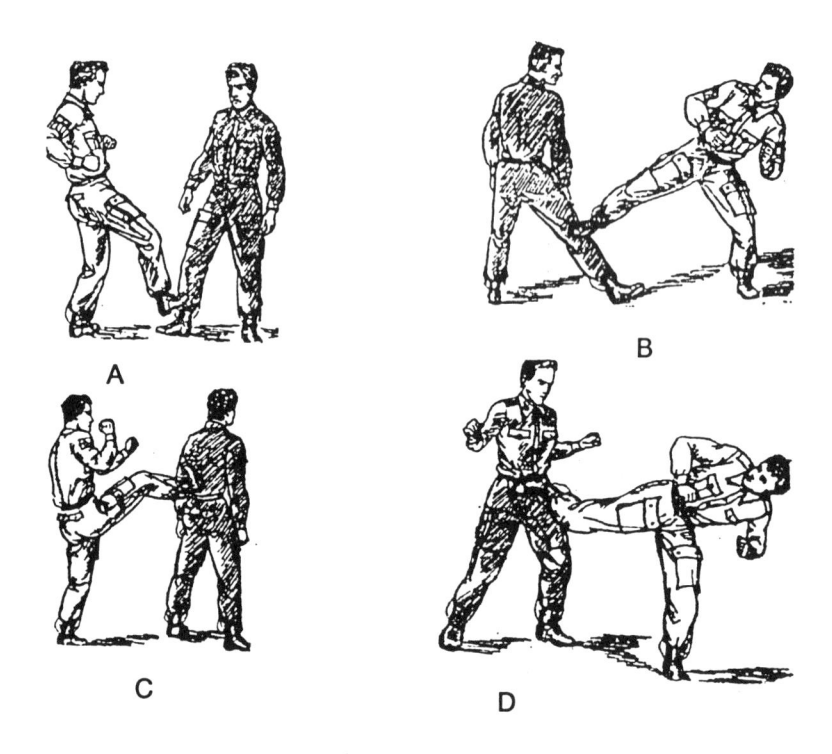

FIGURE 245. DELIVERING KICKS
A—UPWARD
B—TO THE KNEE
C—TO THE BACK
D—BACKWARD TO THE STOMACH

After throwing him by yanking his legs from behind, straddle his back, pressing your knees close to his torso. Slip your arms under his shoulders, get your hands up onto the back of his head, and push down (Figure 254).

Another way of damaging an opponent's neck vertebrae is to take the same straddling position over his back, with one of your feet flat on the ground and the knee of the other leg next to him. With both hands, take him under the chin, pull his head back, and move your upright knee sharply forward toward the side of his head (Figure 255).

FIGURE 246. "GRAIN SACK" THROW

FIGURE 247. OVER-THE-BACK THROW, AND INJURING THE ARM

FIGURE 248. THROW TURNING AN ASSAILANT UPSIDE DOWN

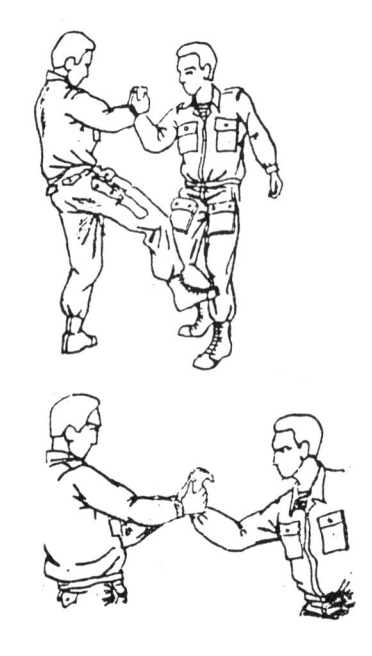

FIGURE 249. PERFORMING AN OUTWARD HAND TWIST

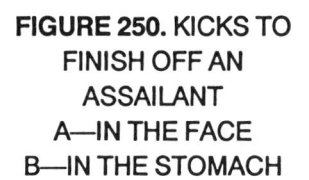

FIGURE 250. KICKS TO FINISH OFF AN ASSAILANT
A—IN THE FACE
B—IN THE STOMACH

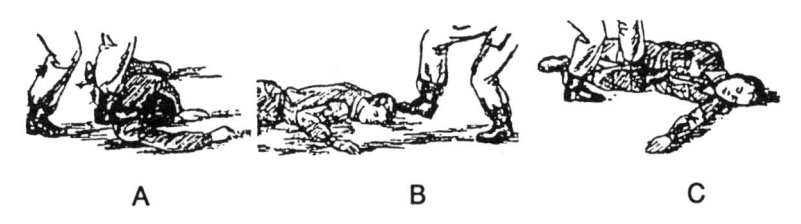

ESCAPING FROM AND FIGHTING OFF PHYSICAL ATTACKS; MUTUAL AID; THROWS

FIGURE 251. KICKS TO FINISH OFF AN ASSAILANT
A—TO BACK OF HEAD
B—TO THE CROWN
C—TO THE CHEST

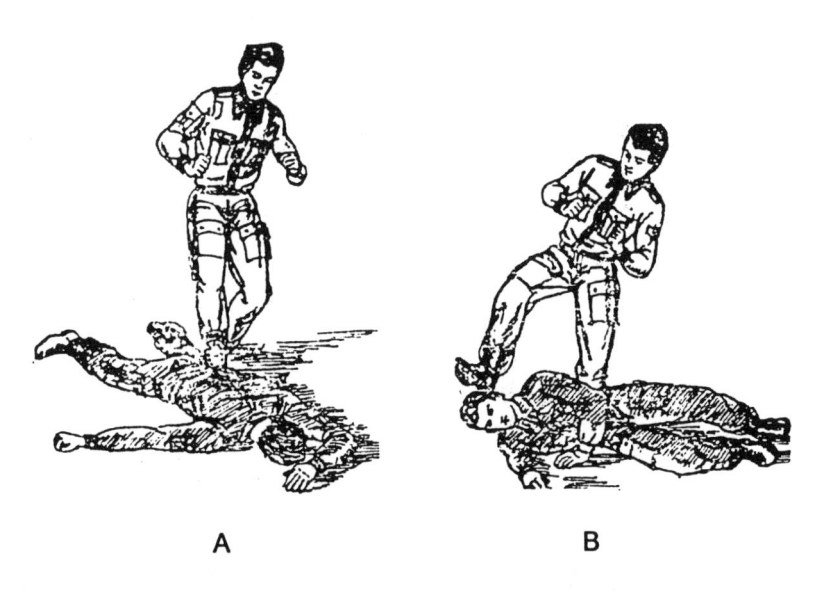

FIGURE 252. STOMPING FROM ABOVE
A—TO THE SMALL OF THE BACK
B—TO THE HEAD

TWISTING THE NECK VERTEBRAE

To twist the neck vertebrae, you have to grab the crown of your enemy's head (by the hair) with the left hand and use the right to take hold of his chin (Figure 256). To deceive your opponent, first twist his

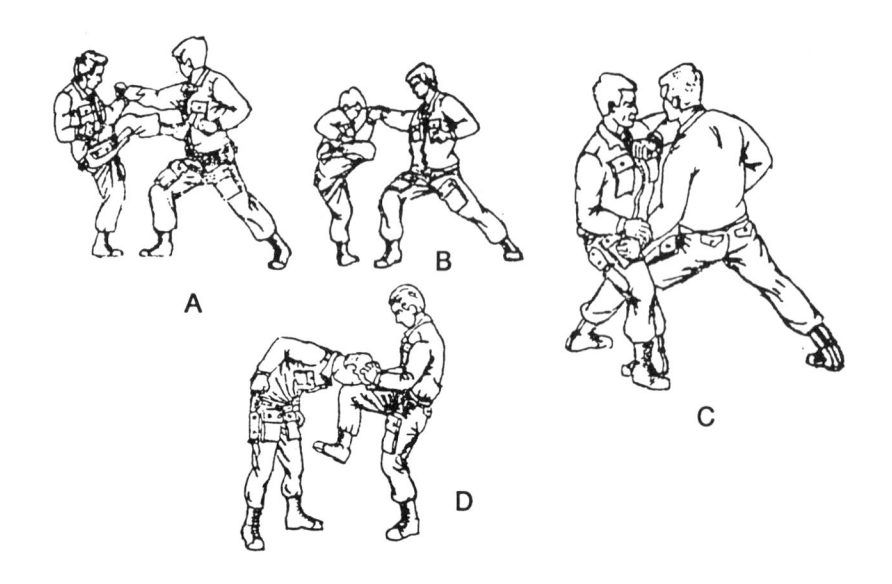

FIGURE 253. BLOWS TO VULNERABLE SPOTS
A, B—KICK TO THE STOMACH
C—ELBOW TO THE FACE
D—KNEE TO THE FACE

head gently to the right, and then sharply twist it down and to the left with your left hand. At the same time, pull his chin upward with your right hand (Figure 256b).

If an assailant gets you down on the ground on your back, maneuver beneath him to get a scissors hold on his body with your feet crossed above him. Grab the hair on the crown of his head with your left hand and get your right hand on his chin, as in the preceding technique. Trick him by lightly twisting his head to the right, then sharply

FIGURE 254. GRASPING THE HEAD TO RUPTURE VERTEBRAE CARTILAGE

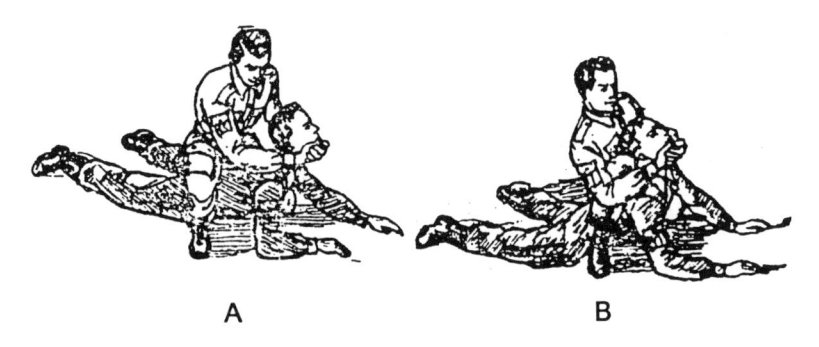

FIGURE 255. BREAKING THE NECK VERTEBRAE
A—THE HOLD
B—THE EXECUTION

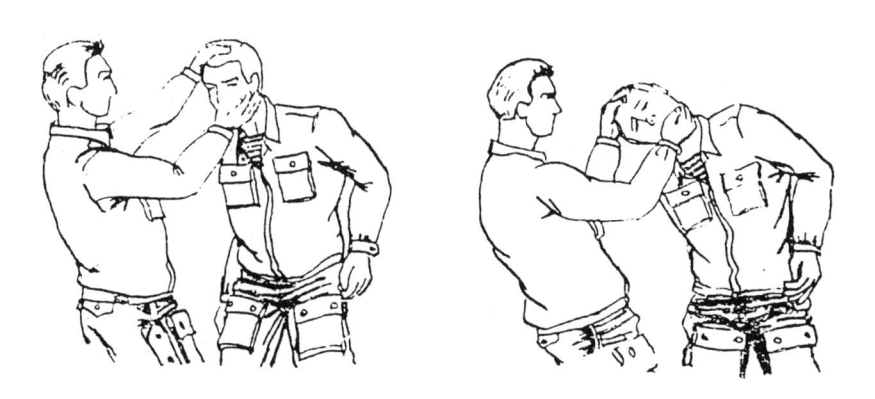

FIGURE 256. TWISTING THE NECK VERTEBRAE
A—HEAD GRIP
B—EXECUTION

yank it toward you by the hair, pushing away hard on his chin. Practice this only on a combat dummy!

CHOKING TECHNIQUES

Choking can be used as another way to beat an enemy and can be done with the help of rope, wire, a stick, a twist of the clothing, hands, and legs. Some of this was examined in Chapter

4. Supplemental to this are simple, combination methods.

Studying and practicing these with your comrades only improves your speed and accuracy in performance. Along with this, you should learn to use your full strength. Combat dummies that are built from bicycle tires filled and arranged to simulate the human body may be used. Motorcycle or light vehicle inner tubes can also be used. The grasping strength of the fingers can be developed in training exercises, as well as the strength of the arm and leg muscles. To be able to defend yourself against chokeholds, the neck muscles in particular should be built up by regular and combat-oriented exercising. There are simple techniques to put pressure on the esophagus or carotid artery; additionally, there are those for the lungs and diaphragm. Chokeholds can be applied to an enemy from a position in front or behind him, when sitting, lying prone or supine, or when on all fours. In any event, the enemy's position is all important in making utilization of choking methods harder or easier. One

A   B

FIGURE 257. HOLD AND CHOKING
BLOW WITH THE FOREARM
A—BLOW TO THE BACK OF THE KNEE
B—CHOKING

thing, however, is decisive: if it is at all possible, get an enemy down on the ground before applying a chokehold.

A forearm blow to the throat and a chokehold with the arms is done thusly: approach your target from behind, grab his hair, eyes, or chin with your left hand and simultaneously push into the back of his right knee with the outside edge of your right foot (Figure 257). While pulling his head back, deliver a sharp blow to his thyroid cartilage (Adam's apple/windpipe) with the inside bone (radius) of your right forearm.

Immediately after this, carry out the chokehold, using the shoulder (his) and the forearm (yours). To strengthen the hold, put your left hand on the back of his head.

The hold is done this way: come up to your target from behind and slip your left arm under his left arm, putting your hand on the back of his head. Put your right arm across his right shoulder and reach across his throat to grab his left collar. Pull it tightly against his throat and push hard on the back of his head (Figure 258).

The following double choking/suffocating technique is done in much the same way as others described previously, but by pressing the enemy to yourself and pushing against the back of his knee to bring him down (Figure 259). Without relaxing your grip on his head and neck, get him in a scissors hold, crossing your feet in front of his stomach. Continue applying pressure on the back of his head and tightening his collar against his throat. Simultaneously squeeze his torso hard with your upper legs. Continue the pressure on his throat

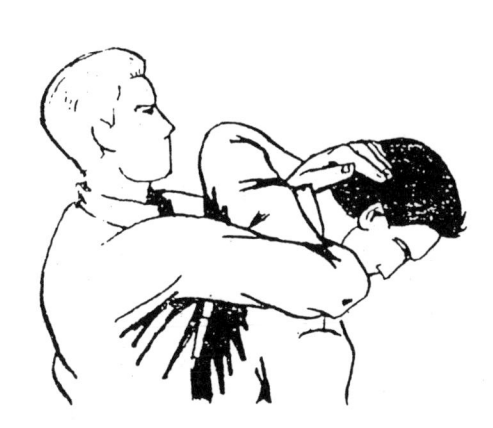

FIGURE 258. CHOKEHOLD FROM BEHIND

FIGURE 259. DOUBLE CHOKING/SUFFOCATING
A—BEGINNING THE MANEUVER
B—CHOKING

with either his collar or your forearm (Figure 259b).

FIGURE 260. CHOKING AN ENEMY ON ALL FOURS

Choking/suffocating an enemy on all fours uses the same grip on the back of his head, and you stand to his left so that your right thigh is against his left side. Put your right arm forward over his shoulder and reach across his throat to grasp his left collar, pulling it hard toward you and up against his throat. At the same time, push hard on the back of his head, moving it forward and down (Figure 260).

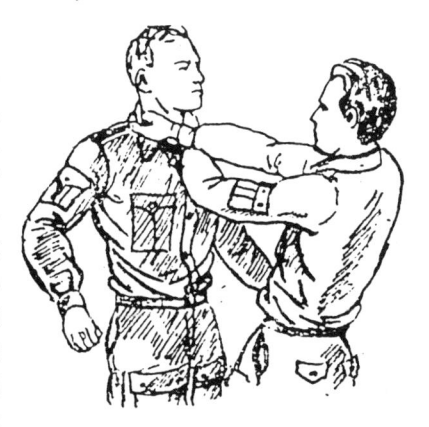

FIGURE 261. CHOKING WITH KNUCKLE PRESSURE

Choking an enemy with his own collar, using a cross-handed grip, is shown in Figure 261. Your knuckles are positioned to press heavily against his carotid arteries. A good, simple, and efficient variation on this is to take his collar with a cross-handed grip and position some of your knuckles to push in on his esophagus. This is especially simple and effective when your antagonist is backed up against a wall or any other barrier.

Choking an enemy with his own clothing and your forearm is described here.

Variant A

Grasp the left front of your target's collar with your left hand, and the right front with your right, pulling the collar as tightly as possible against the nape of his neck. Get your right thumb under the collar (on the outside) and have your four knuckles upward in a good grip. With your left hand, pull downward and toward yourself, pressing in heavily on the throat with the outside bone (ulna) of your right forearm (Figure 262).

The same technique can be used on an enemy lying on his back (Figure 263).

FIGURE 262. CHOKING, USING THE COLLAR

Variant B

When an attacker has gotten you down on your back and is straddling you from on top, counter with the variant A chokehold and get a scissors hold on his torso. Cross your ankles and apply pressure to his chest and middle. Use your hands to choke at his throat and neck (Figure 264).

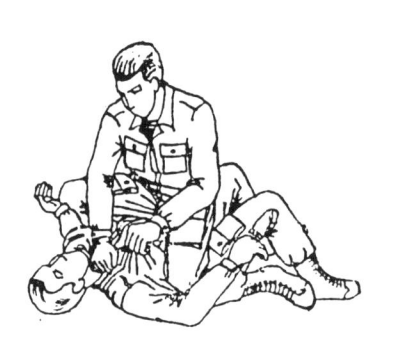

FIGURE 263. CHOKING, USING THE COLLAR

FIGURE 264. CHOKING, USING TWO HANDS ON THE THROAT

USING WEAPONS AND OTHER OBJECTS FOR SELF-DEFENSE

A cold weapon itself can serve as an object to aid in self-defense. Either the blade or blunt part of an ax head can be used to hit, for example, the head, knee, or shoulder (Figure 265).

It can be thrown at a fleeing enemy to kill him silently.

Firearms can be used as supplementary objects to aid in self-defense. For instance, a pistol can still be employed even if it misfires and a foe is coming at you. Hit him in the eye of the

ESCAPING FROM AND FIGHTING OFF PHYSICAL ATTACKS;
MUTUAL AID; THROWS

FIGURE 265. AX BLOW TO THE HEAD

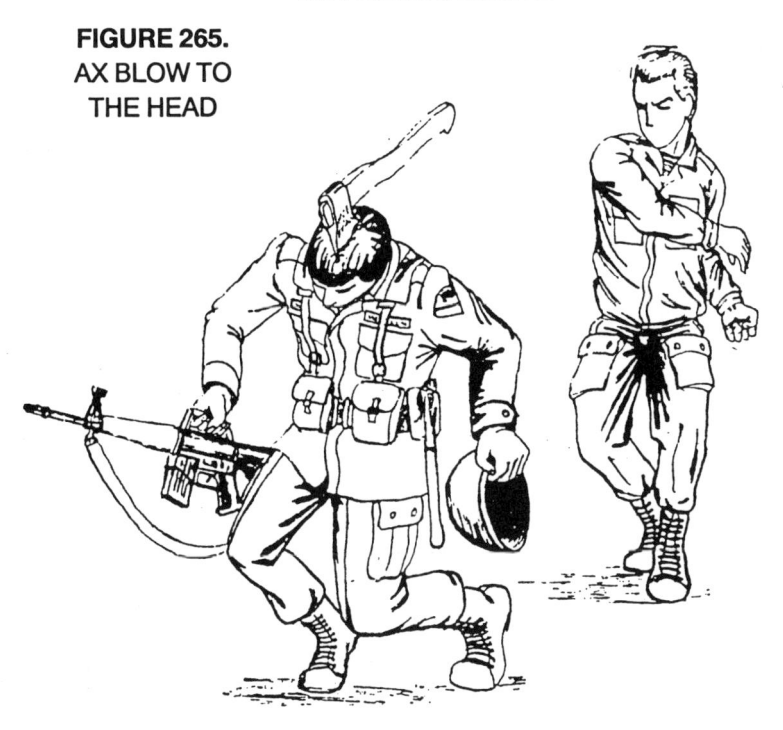

A B

FIGURE 266. A BLOW WITH A PISTOL
A—IN THE EYE
B—TO THE KNEE

KGB ALPHA TEAM TRAINING MANUAL

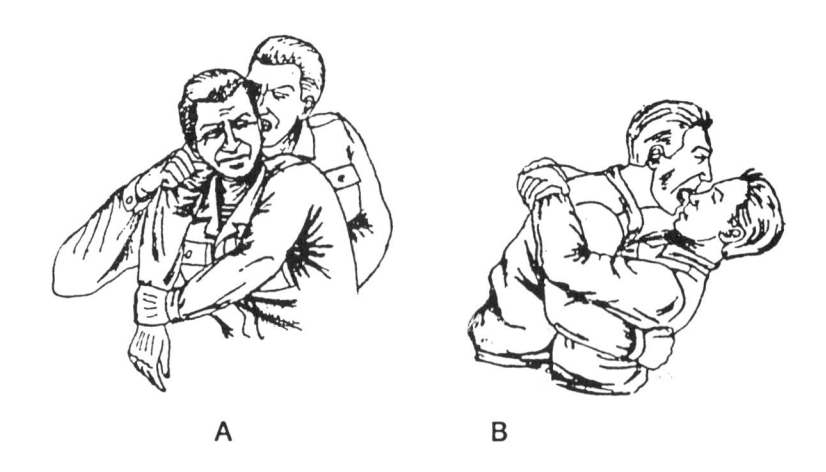

FIGURE 267. BITING AN ENEMY DURING CLOSE COMBAT
A—THE EAR
B—THE NOSE

FIGURE 268. A CHAIR IN THE FACE

side or the knee with a short, slashing blow with the barrel (Figure 266). Before hitting him, use your left hand to feint at his face, drawing attention from the gun.

The teeth serve very well in both defense and attack, when you are in a close hand-to-hand fight, either standing up or on the ground. Whether grappling with an enemy or freeing yourself from his hold, you can "lock" your teeth on his nose, ear, throat, hand, or other part (Figure 267).

Escape from an attack or a hold by using a blow or a throw to get your foe on the ground so that you can use some physical object lying within reach.

In living quarters, you can choose from a wealth of objects to use against an antagonist if you don't have your firearm. It is not at all complicated to use a chair/stool. After hitting him in the face with the chair, you can kick him in the shin, knee, or crotch (Figure 268).

Extraneous, supplemental objects that can be used as weapons of defense are those that are often found anywhere or that can be gotten in hand in a personal combat situation.

THROWING COLD WEAPONS AT A TARGET

The material presented in this section describes the possibilities for employing cold weapons in noncontact single combat. Soldiers can use, for the most part, basic cold weapons for throwing at a target: knives, bayonets, axes, and some pointed or edged items and pieces of equipment. The penetrating (killing) force of a knife thrown is almost doubled. Experiments have shown that a stab down into a board goes in about sixteen to seventeen millimeters, and a thrown knife penetrates to about twenty-seven to twenty-eight millimeters.

Any sharp object can be used for throwing: a nail, razor blade, knife, ax, or any pointed item. Everything that cleaves, cuts, or chops, in the hands of a skilled soldier, should serve a need to destroy an enemy in either contact or noncontact fighting. Figure 269 shows cold weapons available to the personnel of armed forces units, including even a piece of

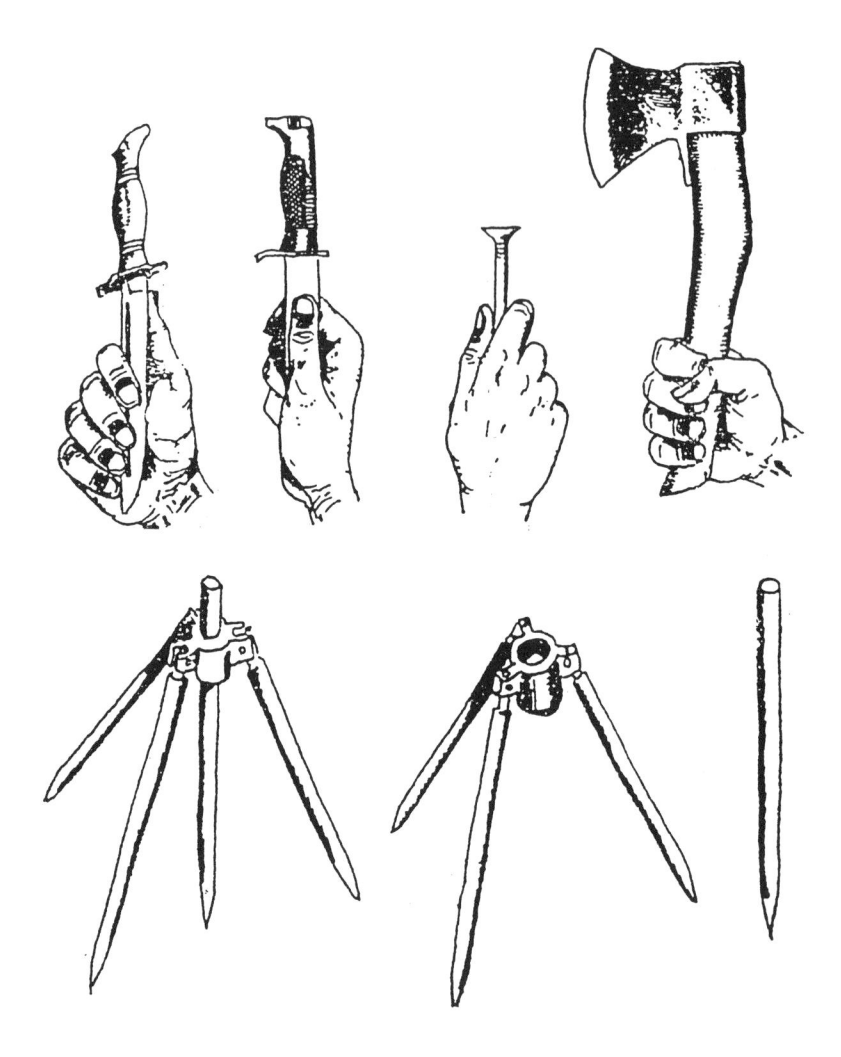

FIGURE 269. COLD WEAPONS

engineer's equipment.

Gripping a weapon for throwing is most easily done by holding its blade or handle. A cold weapon should be thrown from behind the head, like a grenade. In throwing, the following parts of the body should come into play in the sequence shown:

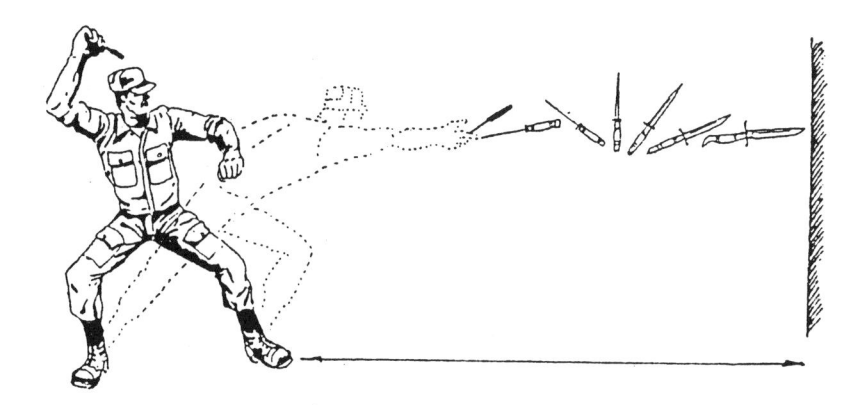

FIGURE 270. THROWING A KNIFE AT A TARGET

FIGURE 271. THROWING AN AX AT A TARGET

feet, back, shoulder, arm. The throwing hand performs the decisive step, with the opposite foot forward. Throwing from behind the head produces the most powerful throw. The throw's distance depends on how high up you hold the blade (of a knife)—with the sharp edge away from you.

If an enemy is moving toward you, the knife is thrown with a grip closer to the tip of the blade. Conversely, to hit a target at a greater distance, the grip has to be placed farther up the blade, that is, closer to the hilt. The basic combat distance is five to six paces, or closer, to ensure hitting the enemy. He should be hit before he has a chance to duck, twist, dodge, fall, or hide. It is possible to hit a target at ten paces, but that demands an exact calculation of the weapon's revolutions in its flight (in which time the target can get to shelter or avoid being hit). For combat throws, the weapon should make one full half-turn, so that the point strikes (Figure 270).

An ax thrown at a target should make one full turn (Figure 271). If you are throwing it at a stationary target, you can double the throw's distance so that it makes two full turns.[1]

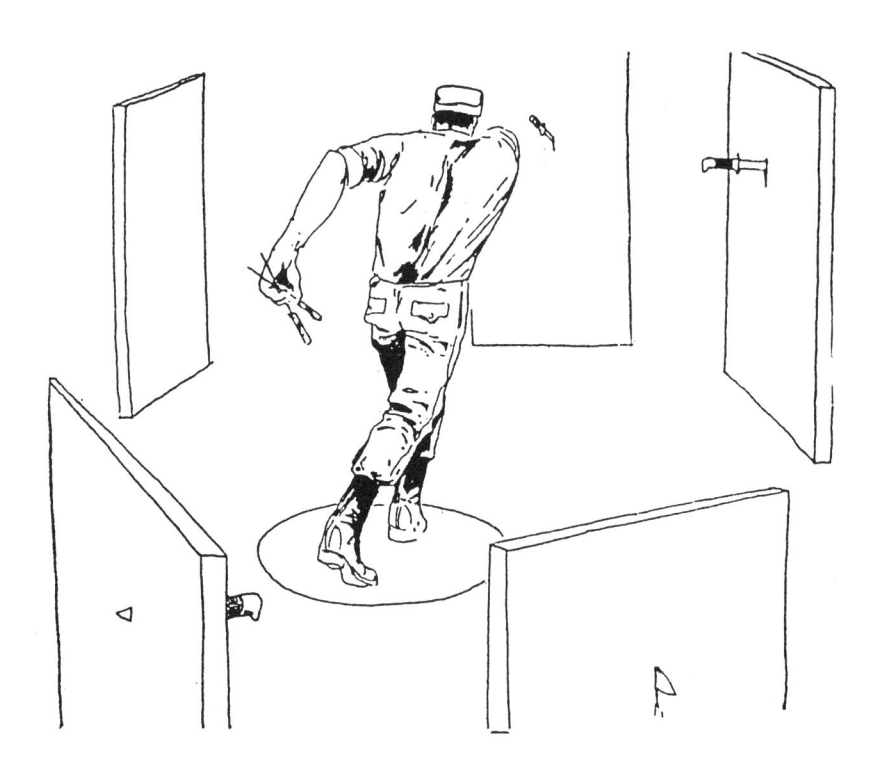

FIGURE 274. THROWING AT TARGETS IN A CIRCLE

The Technique of Throwing

Grasp the cold weapon (knife) by its blade and, with a sweep from behind your head, with the free hand out forward, throw the knife at the target. Let the knife go at eye level, aiming for the chest of your target (who is standing at full height). A crucial element of the technique is guiding the weapon at the decisive moment of its leaving your hand. You have to extend yourself along and into the vector of the throw. This fundamental part of the technique has to be developed through repeated practice with and without a weapon at the beginning of throwing training. To preserve and maintain serious weapons, begin the learning

FIGURE 275. USING A RAZOR BLADE
AS A THROWING WEAPON
A—THROWING
B—HOLDING THE BLADE

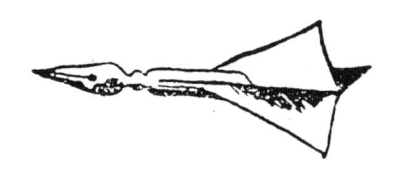

FIGURE 276. USING A PEN POINT FOR THROWING

process by using spikes or big nails. Keep some insulators at hand to incrementally increase the weight of the practice weapons: a weapon made heavier travels faster and hits harder.

In independent practice, you should work under someone with more experience, preferably a training unit leader.

Other sharp or pointed objects may be used for throwing as a means of defense. Things that can be found in an office or a desk can be employed (e.g., razor blades, pen points, needles) to fight back, gain time, or act as follow-up, effective self-defense measures.

NOTES

[1] Text was missing in original manuscript. Figures 272 and 273 were also missing.

CHAPTER 6
PENETRATING BUILDINGS IN AN ATTACK

To carry out operations inside a building, the first concern after scouting and surveillance, is how to take out any sentry. A team of no more than two individuals should be detailed to kill him silently. The pair moves quietly, creeping or making short rushes along the walls, passing under windows in a crouch, and stealing up on the sentry (Figures 277 and 278). Waiting for the right moment, they move to attack from the rear, out from behind a corner, and kill the sentry.[1]

Once he has gotten inside the building, he throws the rope down, securing it at his end, and the others climb up it.

Getting inside a building can also be done with the help of a rope and grappling iron ("cat") tied to one end, which is hooked into a window (frame) or onto a balcony (Figure 284).

Using a rope suspended from a helicopter, a team can get on a roof, from which (using access doors or windows), it can get into the building. Once inside, movement has to be done carefully but quickly and decisively—staying close to walls, avoiding long stays by windows or door. Inside, at the climactic stage of a capture or raid, it can boldly employ assault rifles, pistols, or grenades: that is, at doorways and in offices, the appropriate means is selected. During all this, it is best to get right up to the sides of entranceways, with team members on both sides. Entrances must be made immediately after firing in

FIGURE 277. CONCEALED MOVEMENT ALONG A BUILDING

or tossing in grenades (Figure 286). In any living quarters, to deal with the unexpected, all should know how to direct fire and use bayonets or butts, and how to close in for hand-to-hand fighting with knives, fists, feet, teeth, and so on (Figures 287 through 289).[2]

In an urban setting, an enemy going along a street can be taken, using an automobile. You should drive up behind him quietly, and two team members have to jump quickly out of the car, one to get in front of, and the other behind, the target. A choke hold is used, as well as something to throw over his head. Get him into the back seat of the vehicle, holding on to his neck and feet. During the capture, you can hit him a stunning blow to the head with some kind of sap/blackjack. These operations can also be done by jumping out from behind a corner or by taking advantage of a bend in the street (Figure 291).

Capturing an enemy in the woods is done by three persons. At a bend in a road, a "sudden obstacle" should be built (something a driver sees at the last moment and where he cannot stop the vehicle in time to avoid it). After the vehicle is halted, one team member

PENETRATING BUILDINGS IN AN ATTACK

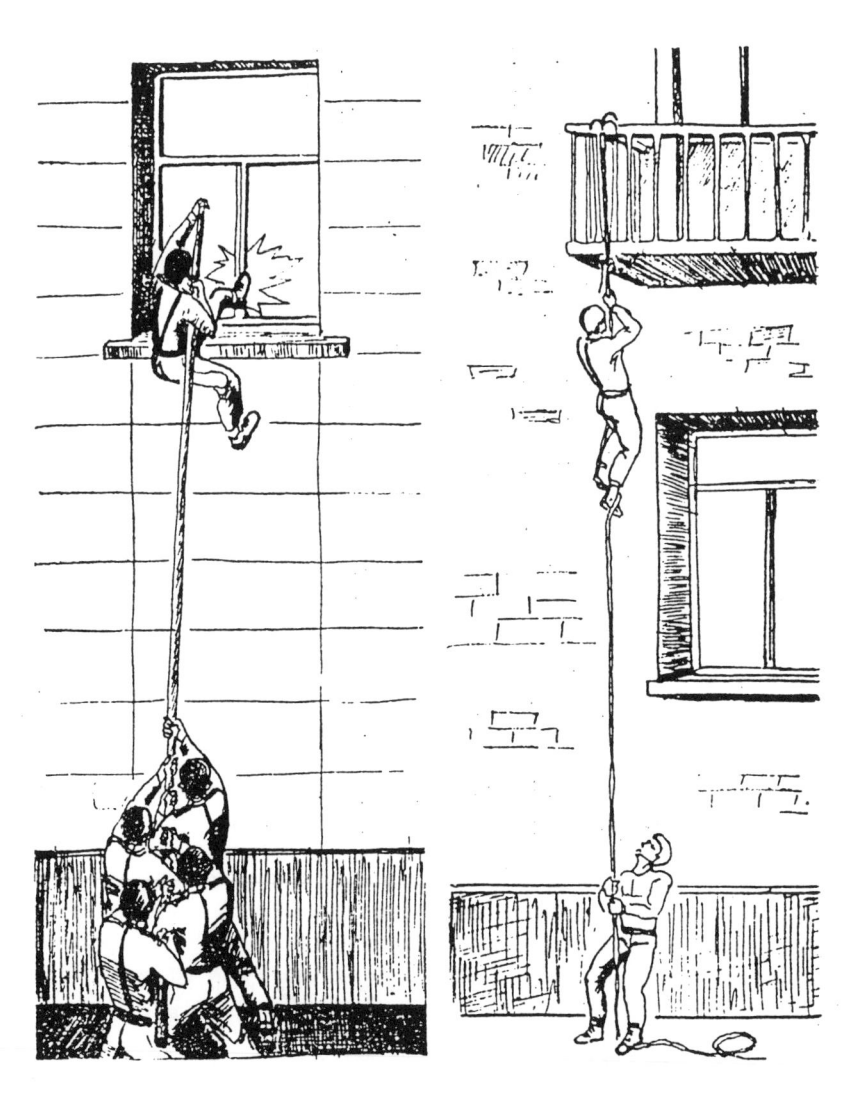

FIGURE 283. STORMING A BUILDING USING A POLE

FIGURE 284. STORMING A BUILDING, USING A GRAPPLING IRON

kills the driver and the other two, using choking techniques, pull the passenger out of the front seat and put him in the back. The first team member drives the car off (Figure 292).

Another way to get a captured enemy into a vehicle is by holding him by his hair and crotch from behind.

Taking an enemy who is sitting behind a table in a house or apartment is done as follows. Come right up to him and knock him over along with the chair he is on, which makes it harder for him to get up. Hit him on the head with a pistol and twist his arms behind his back (Figure 293).[3]

NOTES

[1] Text was missing in original. Figures 279–283 were also missing.
[2] Text was missing in original. Figures 286–289 were also missing.
[3] Text was missing in original. Figures 292–294 were also missing.

FIGURE 285. LANDING A TEAM ON A ROOF FROM A HELICOPTER

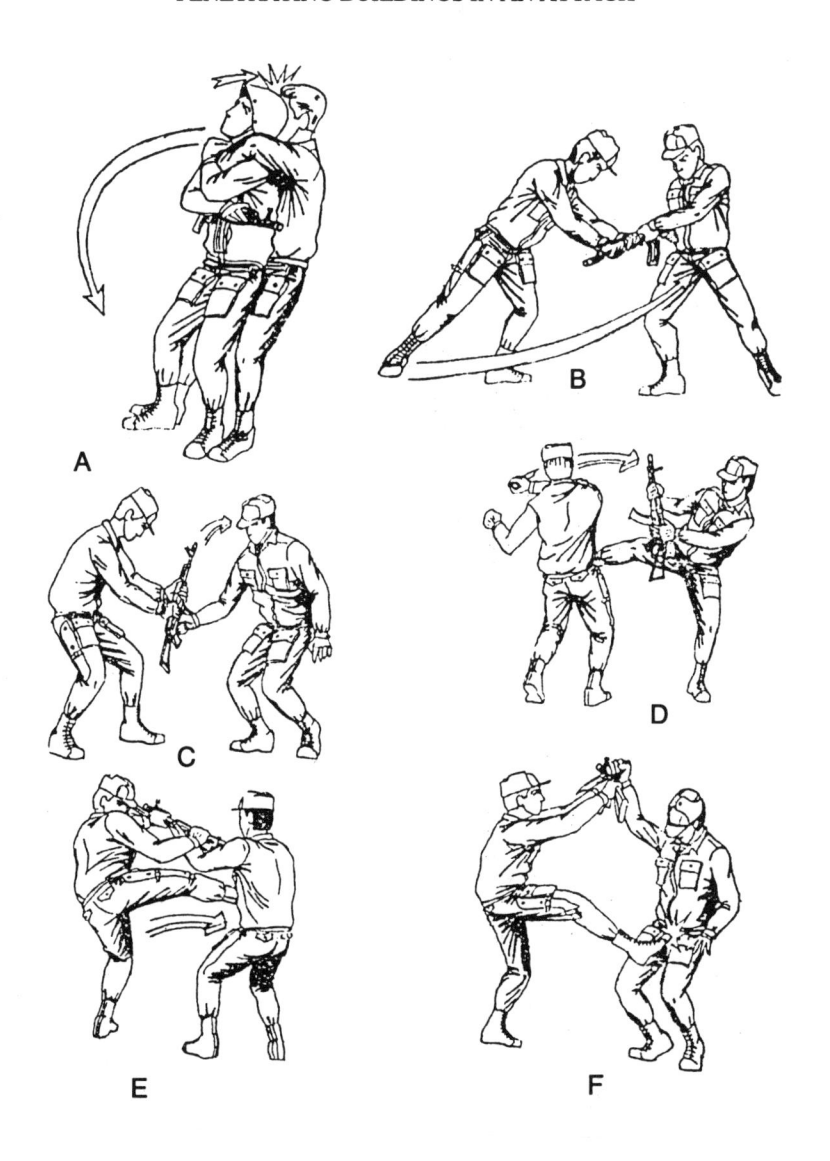

FIGURE 290. PERSONAL COMBAT IN CLOSE QUARTERS
A—BREAKING OUT OF A HOLD
B—DEFENSE AGAINST A JAB WITH A RIFLE (OR GETTING SHOT)
C,D,E,F—DEFENSE AGAINST COLD WEAPON ATTACKS

KGB ALPHA TEAM TRAINING MANUAL

FIGURE 291. CAPTURING AN ENEMY IN A TOWN

FIGURE 295. CHARGING DEFENSE AGAINST AN AX BLOW (WITH SHOULDER OR HEAD) INTO THE STOMACH

PENETRATING BUILDINGS IN AN ATTACK

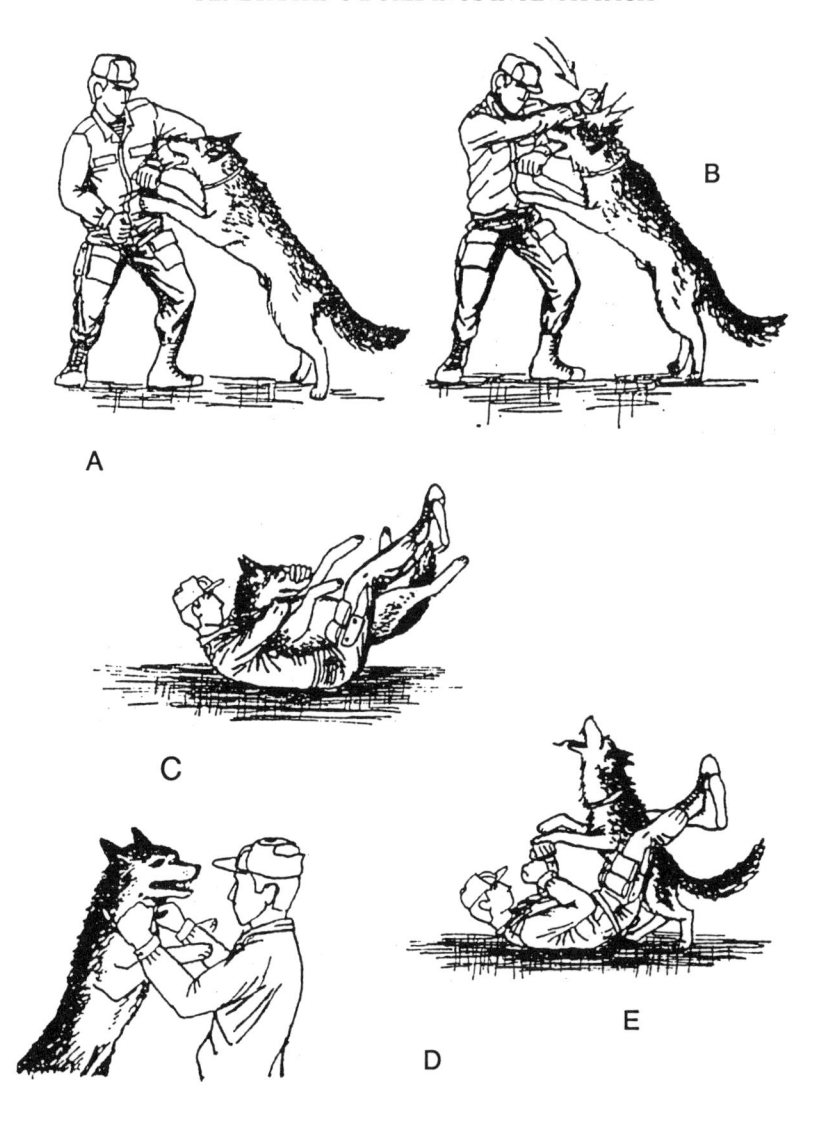

FIGURE 296. METHODS OF DEFENSE AGAINST A DOG
A—JAMMING IN THE FOREARM
B—HITTING THE HEAD WITH A FIST
C—CHOKING WITH THE HANDS/ARMS
D—CHOKING WITH THE DOG'S COLLAR
E—USING A SCISSORS HOLD

CHAPTER 7
MODELS FOR RESTORING WORK CAPACITY AND MONITORING THE STATE OF HEALTH

Physical strain and restoration of work capacity is a dialectical process. Restoration is possible only when energy expenditure is made up for fully in a restorative process.

Restoration of work capacity is not just the returning of physiological functions to an operation level, but the putting in place of a system of new, higher level energy potential. Depending on the extent of strain to the system, full restoration may take four to seven days.

TYPES OF PHYSICAL RESTORATION

Intensive physical activity without damage to the health is possible only when using a system of restoration (medical-biological, psychological, and pedagogical methods). In this, the pedagogical system can be regarded as fundamental and primary because it must determine the regimen for each trainee and combine the work loads and rest at all stages of his training.

Pedagogical System of Restoration

This includes:
• Rational planning of heavy work loads in the training curriculum for the week and month, by courses and objectives
• A correct mix of work and rest in the course of a day

• Introduction of special exercises for weakened muscles at the end of each work session, by discipline, individual morning exercising, establishing a positive emotional foundation, and so on

Medical-Biological System of Restoration

This includes:
• Correct assessments of individuals' health
• Proper nutrition during the workday and week
• Utilization of pharmacological means (especially vitamins)
• Physiotherapeutic and balneological methods (e.g., steam baths and different kinds of massage and self-massage)

Psychological System of Restoration

This is used in combination with the pedagogical and medical-biological systems for ameliorating neuropsychological tension (i.e., psychotherapy). This system's methods include counseling, leisure time for personal development, cinema, concerts, and any interesting changes in routine. Especially valuable in the case of heavy physical exertion and work loads are steam baths followed by cold showers, contrasting-temperature showers, and quiet soaks in warm water.

Contrasting showers of five to seven minutes' duration can be a simple, effective means of restoration, with the following methods of application: one minute of hot water (+38 degrees C) then five to ten seconds of cold water (+12 to 15 degrees C), and so on.

REHABILITATION

This restores the health and work capacity of a trainee after an illness, injury, or pathological condition. The rehabilitation process employs methods and means discussed above, as well as lighter, clinically determined exercises.

Massage and Self-Massage

These are used after physical strain to rebuild more quickly an individual's work capacity. Restorative massage (RM) can begin within twenty to thirty minutes after exertion, and can go for seven to twelve minutes. Massage relaxes the muscles and parts of the body affected by heavy strain, as does kneading, stretching, and contracting the extremities.

In cases of significant fatigue, RM can be done within one to two hours after the exertion, for about fifteen to twenty minutes; for extreme fatigue, within three to four hours afterward, for about twenty to thirty minutes.

Massage is performed lightly at first, then more deeply and energetically, using a lubricant (e.g., soap, Vaseline).

Self-Massage

This can be done by the individual trainee, and he can utilize various kinds of powder, lubricants, or lotions. In order to prevent skin irritations and to facilitate massage, boric Vaseline and olive or mineral oil may be used. After massage, Vaseline can be removed from the skin with gauze pads moistened with warm water or eau de cologne. For this and in baths, regular hand soap should be used. Massage action during self-massage is the same as for administered massage. The sequence is as follows: relaxing and working on the leg muscles, the muscles of the back, arms, head and neck, the front of the legs, the stomach, and the chest.

Contraindications for Massage and Self-Massage

Massage of either kind is not done when there are suppurations or inflammations, skin diseases, damage to the skin surface, body temperatures above 37 degrees C, varicose veins, or noticeable fatigue right after physical exertion.

Massaging concentrations of lymph nodes in body cavities (e.g., behind the knees or under the arms) is never done.

STEAM BATHS

The bath is not only a hygienic device, but also plays an important role in restoring work capacity. In addition, it is used for regulating and getting rid of excess weight. Along with the steam bath, the dry-air bath (sauna) and the portable warm-water bath, "Termika," are used. In steam baths the air temperature is at 60 to 70 degrees C, and the relative humidity is in the range of 20 to 70 percent. In the sauna the air temperature reaches 100 to 140 degrees C, with a relative humidity of about 10 percent. With daily visits to the steam bath, the human system adapts to the effects of high temperatures and the humid air, and to a marked fluctuation in salinity change. Perspiration in high temperatures followed by cooling off in another room (18 to 20 degrees C) promotes elastic activity in the vascular system, flushes the skin pores, and lightens the filtering work of the kidneys—promoting, so to speak, "rejuvenation" and hardiness of the whole system.

The way to use the steam bath is given here.

After arriving at the steam bath (which has first been cleaned with a disinfectant), it is best to take a warm shower (35 to 40 degrees) first, without getting the head wet. Before entering the steam room, a felt hat is put on (or a hat with a wide brim). The first four minutes in the steam room are spent in warming up at floor level, followed by five to seven minutes on a higher-up shelf (depending on temperature, humidity, and comfort). It is better to start the steaming process lying on the stomach, with the ready help of a comrade or masseur nearby: under the room's conditions the muscles become quite weak.

It is very useful to employ a switch (bundle) of birch or other twigs with thick foliate (a dry switch is first exposed to the steam). Application of the switch should move from the heels up to the buttocks, then farther up the back and out along the upper extremities, and from them along the sides and flanks. After three or four complete series of this, continue using the switch on yourself in combination with steaming. To do this, hold the switch up, raising it as high as possible in the steam chamber to heat it to the same temperature as the air. With gentle movements, slowly press it to the body. Hold it up again and press it some more, very briefly, to the body, using the highest (hotter) end of the foliage. In the same way, press the heated leaves of the branches to, first, the back of the legs and back, lower back and buttocks, and flanks and lower legs. (Do not do

MODELS FOR RESTORING WORK CAPACITY AND MONITORING THE STATE OF HEALTH

this to the backs of the knees, or expose them to the steam!) During subsequent visits (no more than four times in any series), use the switch to rub during steaming.

Beginners may take the steaming one time—but no more than two times—for six to nine minutes.

For the cardiovascular system and all of the skin surface, it is beneficial to take steam baths, using different techniques: e.g., using two switches, one of which is dipped in cold water in a nearby basin. This switch should be held to the body in the first phase of steaming, but used only for light whipping on the body in the second phase.

Loss of excess weight does not happen right away by using a steam bath. But, to this end, spend three to four minutes at the lower level of the steam room, periodically wiping the sweat off the upper body with a dry towel or with a scraper (the edge of a soap dish will do). Upon leaving the steam room, rinse off in a warm shower, best using soap and a handful of bast, and not getting chilled. You should dry yourself and wrap up in a sheet or warm (knit) clothing, leaving the head free so you can breathe. Lie down and rest for fifteen to twenty-five minutes. Then dry yourself completely and repeat the whole procedure two or three times.

The bath as a restorative means after working heavily and to one's limit, with a following day for rest, is used in a particular way. The individual enters the steam room three to four times for five-to-nine-minute sessions, into a temperature of 110 to 120 degrees C (plus or minus 10 degrees C). After each time, a brief cooling shower or dip of 10 to 15 seconds' duration is recommended, followed with a warm shower of up to three minutes. This procedure increases cardiac activity and throws off strain and fatigue. The breaks between steam room sessions are about six to ten minutes. The warm-water temperature should be 30 to 40 degrees C.

If the recommendations are not followed, sickness or extreme fatigue can be brought on by overheating. Symptoms in the first stage are, characteristically, agitation, nausea, dizziness, headache, and frequent need to urinate, which are followed by sluggishness, shortness of breath, increased salivation, and a complete halt to perspiration. When these signs appear, the affected individual should be taken to another room, wrapped warmly, and put by an adequate fresh air source. Then he should be given ammonium carbonate to breathe and hot sweetened tea with lemon to drink. Alcoholic (spirits) drinks are categorically avoided in the recovery period; only sweetened tea or fruit juices can be given. The question of the individual's returning to the steam bath is decided only after consultation with a physician.

At the present time, domestic industry has begun the manufacture of a portable warm bath under the label "Termika." It consists of two components. One contains a heating device; the other a heat compartment (a heat-resistant tent made of two layers of nylon with a foam lining between them). Unaffected by the raising of the air temperature (up to 130 degrees C), an individual experiences no difficulty in breathing. He breathes room-temperature air, just as if his head were outside of the tent. Sessions of up to thirty, forty-five, or sixty minutes at temperatures of up to 80 degrees C, in combination with self-massage, fully conform to the goal of restoration after heavy physical strain. a regimen for a weight loss of two to three kilograms per session is three fifteen-minute breaks.

NUTRITION IN TIMES OF HEAVY PHYSICAL EXERTION

To maintain normal physical functionality, trainees in times of heavy physical exertion absolutely need rational nutrition (lots of calories, variety, ease in digestion, and moderate daily portions). Each trainee should know the approximate nutritional parameters. A proper diet covers energy expenditure, increases work capacity, and speeds up the restorative processes of the body. To establish proper nutrition, selection of foodstuffs, their chemical content, proportions of different foods in the rations, preparation methods, dietary issues, and so on have to be known.

In working out a course of nutrition, it is of primary concern that there be knowledge of the biochemical mix in a selection of foods,

MODELS FOR RESTORING WORK CAPACITY AND MONITORING THE STATE OF HEALTH

TABLE 4

	PROTEIN, GRAMS	FATS, GRAMS	CARBO-HYDRATES, GRAMS	CALORIE CONTENT, KILOCALORIES (KC)
VARIANT 1 (WINTER)	2-2.3	2-2.1	10-11	60-72
VARIANT 2 (SPRING-SUMMER)	2.4-2.5	1.7-1.8	9-10	65-70

basic portions, the intensity and character of physical exertion, and the process of absorption by a healthy person's system.

The fundamental demand of nutrition is that its energy value correspond to the system's (energy) expenditure. Considering the incomplete digestion of some foods, the calorie content, then, has to be increased by 5 to 10 percent for daily energy expenditure. The average calorie content and amounts of albumen (protein), fats, and carbohydrates in the rations are determined by the extent of energy expenditure. From participation in field training exercises, marches, forced marches, and other such activities, it is known what the theoretical contents of rations are for one kilogram of (an individual's) body weight. These are given in Table 4.

The relevant energy values of foods are the collective kilocalories (Kc) given for each gram of food for consumption:

$$\begin{aligned} 1 \text{ gram of protein} &= 4.1 \text{ Kc} \\ 1 \text{ gram of fat} &= 9.3 \text{ Kc} \\ 1 \text{ gram of carbohydrate} &= 4.1 \text{ Kc} \end{aligned}$$

The full value of nutrition is determined by the three basic food components, their concentration, and their optimal relationships that facilitate their fullest absorption.

For the training period, it is recommended that the following daily requirements be met: up to 170 grams of protein, up to 160 grams of fat, and up to 650-700 grams of carbohydrates, which form a (theoretical) ratio of 1:0, 8:4; and of the calorie content of the whole (daily) ration protein constitutes 15 percent, fats 25 percent, and carbohydrates 60 percent.

Energy expenditures vary within a norm; therefore, daily requirements, based on type of physical activity, can be broken down for:
- Personnel doing desk work—3,000 Kc
- Personnel performing light physical labor—3,500 to 3,800 Kc
- Personnel engaged in strenuous physical labor—4,500 to 5,000 Kc or more

If a calculation of the calorie content is done, which is essential for safeguarding trainees' physical systems, these are the results:

Protein:	170 g x 4.1 Kc	697 Kc
Fat:	160 g x 9.3 Kc	1,488 Kc
Carbohydrates:	650 g x 4.1 Kc	2,665 Kc
	Daily Total:	4,850

The content of nutritional foods include protein, fats, carbohydrates, minerals, vitamins, and water.

Protein

This is a complex chemical substance that is utilized in building cells, tissue, antibodies, and hormones. It is found in meat and dairy products, fish, eggs, and, to a lesser degree, in potatoes, beans, soybeans, buckwheat, oats, barley, and rice. Almost all of these are rich in so-called lipotropic factors, which block surplus fat concentrations in the liver's cells. The liver's glycogen helps store these substances, and that increases the liver's efficiency. A no less important and different role is played by the lipotropic factors mitionin and choline. Their presence in the blood strengthens the formation of phospholipids (fat-like substances made up of phosphorus and nitrogen), which break up cholesterol deposits in the blood vessels (preventing atherosclerosis).

Fats and Carbohydrates

These are the basic sources of energy. There are both animal and vegetable fats. Animal fats come from butter, sour cream, bacon, mutton, beef, and so on. These fats have the highest nutritional value and promote the absorption of vitamins A, D, and E.

Vegetable fats are from sunflower seeds, maize, nuts, flax seed, and other oil bearers, and constitute the valuable polyunsaturated fatty acids, which lower cholesterol levels and are generally taken as vitamins.

Carbohydrates are found in baked goods, grains, sweets, vegetables, and fruits. These contain a lot of vegetable fiber, without which the stomach and intestines would not maintain their function of regularity. Large amounts of the mineral salts—potassium, magnesium, natrium (sodium), phosphorus, iron—are generally found in our systems.

Potassium

This strengthens the contractions of the heart muscles, increases blood flow, and reduces tissue swelling. It is found in potatoes, rice, cabbage, plums, raisins, and oats.

Magnesium

This is beneficial to the central nervous system and cardiac functions; it reduces strain in the muscle walls of internal organs, reduces cholesterol, and builds up intestinal functionality. Bran, green vegetables, beans, carrots, oats, buckwheat, and wheat are rich in magnesium.

Sodium

This improves nerve function and regulates saline exchange in muscle tissue.

TABLE 5

PHOSPHORUS	CALCIUM	MAGNESIUM
4.0	2.0	0.8

POTASSIUM	IRON	SODIUM
5.0	20.0	UP TO 15-20

In recent years, scientists have ascertained that a healthy person does not need more than five grams of salt a day, even though we are used to an intake of about fifteen grams. This excessive use is particularly harmful to those who suffer from cardiovascular diseases. For these individuals, it is beneficial that they have salt-free milk or artificial milk products, or fruit products on days of light physical exertion. For periods of heavy exertion in training, the daily requirements are given in grams in Table 5.

VITAMINS

Increases in daily vitamin demands require a very extensive variety of foods during a training period, as well as the preparation of foods rich in protein and carbohydrates. Under conditions of regular, scheduled feeding, this requirement is fulfilled by natural foodstuffs. In periods of increased work loads, such full-value, multiple vitamins as "Undevit" or "Dekamevit" can be taken in one or two doses, two or three times a day at meal times. Also, to regulate oxidation during the restorative process, supplementary ascorbic acid can be taken in doses of 100 to 200 milligrams.

Vitamins are an irreplaceable part of foods. A lack of these valuable nutritional elements leads to chronic aggravation of gastrointestinal illness in the winter-spring period, especially in times of increased physical activity. In the early spring, even healthy people sometimes start to notice sluggishness, increased irritability, a quickness to fatigue, and sleeplessness. This deteriorated state of health can be caused by a vitamin deficiency.

Vitamin A

This is needed for normal functioning of the mucous membranes and skin. Vision deteriorates without it. It is found in carrots, apricots, tomatoes, sorrel, raspberries, and dairy products. A recommended daily dose is two to three milligrams.

Vitamin B Group

This is an element in many conversion processes, in the activity of the nervous system, and in the gastrointestinal tract. It is found principally in lean meat, fish, dairy products, black bread, yeast, bran, green vegetables, and legumes. The daily requirement can be met by a yeast extract or drink. The daily vitamin requirements are B1—3 milligrams, B2—3 milligrams, and B3—2.5 milligrams.

Vitamin C (ascorbic acid)

This strengthens the body's resistance to various diseases, lowers blood cholesterol levels, and is good against liver ailments. Foods rich in this vitamin include currants, rose hips, gooseberries, wild and domestic strawberries, oranges, mandarin oranges, lemons, and fresh greens and vegetables. The daily requirement is 150 milligrams. (It is especially effective in rose hip tea.)

Vitamin P (citrin)

This strengthens the blood vessel walls. It is found in tomatoes, citrus fruits, cabbage, lettuce, and parsley. The recommended daily dose is up to 80 milligrams.

Vitamin E (tocopherol)

This is essential for human reproduction. It is found in greens and egg yolks.

WATER

An individual daily requirement of liquids is up to 3 to 3.5 liters, depending on the season and climatic conditions in which physical activity takes place. The human system produces only

TABLE 6

1-2 HOURS	2-3 HOURS	3-4 HOURS	4-5 HOURS
WATER, COCOA, COFFEE, MILK, TEA, BOUILLON, SOFT-BOILED EGG	COFFEE WITH MILK OR CREAM, HARD-BOILED EGG	STEWED CHICKEN OR BEEF, BLACK BREAD, APPLE	ROAST MEAT OR GAME, SALT FISH, GREASE PORRIDGE, BEAN SOUP

100 milliliters of urine from 1 liter of water; the remaining water is evaporated in thermo-regulation. Small drinks of 80 to 100 grams over the course of an hour make for the best absorption.

Periods between meals should not exceed five hours. Data on the length of time the stomach retains food (according to N.N. Yakovlev) are given in Table 6.

All route marches that include heavy physical exertion bring about energy expenditures and require adequate nutrition along the route. With such exertion, dissipation of carbohydrate reserves in the body has to be carefully guarded against; such losses can lead to hypoglycemia (weakness, loss of general working ability, and right up to the inability to move on one's own). Sugar is used to treat this, but the best way to deal with it is with liquid food containing a mixture of vitamins and minerals that is easy to digest. The simplest mixture, which can be made for any person, should contain 50 grams of sugar or glucose, 200 milliliters (a glassful) of fruit or berry juice (or tea with jam), up to 1 gram of ascorbic acid, and 0.8 to 1 gram of sodium (table salt). The simplest drink is tea with jam and lemon. A little lump of sugar should be taken with the drink.

WAYS OF MONITORING THE STATE OF HEALTH

Within the bounds of moderate physical stress, a normal

intensity of physical activity is calculated as when the heart beat reaches 150 to 160 beats per minute. The work required, for instance, to overcome physical obstacles pushes the heartbeat up to 220 beats per minute.

A simple method for checking normal heartbeat rate is an orthostatistical test. In the morning, immediately after waking up, the subject's pulse is taken for thirty seconds while he is still lying down. After he gets out of bed, the pulse is again measured as he stands in place. The difference in pulse rates, of up to thirty beats (the rate while standing is faster), indicates a normal frequency of heartbeat rate. A simple, practical, and sufficiently accurate measurement of normal, functional blood circulation is made with a pulsometer. The pulse rate is taken from the small blood vessels in the fingers, which are held up in the air; from the carotid artery; the radial artery and others; and by auscultating the heart. It is recommended that the pulse of a subject at rest be measured in ten-second segments repeated two or three times in series in order to obtain reliable numbers for normal or disturbed rhythm (arrhythmia).

During normal physical activity, the pulse rate of a person at rest is about sixty to seventy beats per minute.

Monitoring physical activity is tied to measurement of vital lung capacity (VLC). This measuring is done with the spirometer, and it gives an indication of the maximal amount of air that can be exhaled after a maximally deep breath is taken. The size of VLC is affected by the position of the chest cavity, the condition of the respiratory muscles, and the blood capacity of the lungs.

The absolute maximum of VLC fluctuates between 1,800 and 2,700 milliliters. The extent of a full, model exhalation is 1.5 to 2.5 seconds. When bronchial function is obstructed, the exhalation is longer. On average, breathing rate is fourteen to eighteen times per minute, and for athletes it is ten to eleven times per minute. The normal VLC is about 3,000 to 3,500 milliliters.

The method for measuring the length of time a breath can be held after inhalation is as follows. A seated subject takes the deepest breath possible, and, at the fullest extent of the inhalation, he holds his breath in, pinching his nose shut.

In performing the measurement of how long a breath can be held out (exhalation), inhaling and exhaling both must be just average. Holding an inhaled breath is measured at about fifty-five seconds; an exhaled breath lasts about thirty to forty-five seconds.

The most important organs for removing "slag" from the system are the kidneys. Their function can be examined externally. The end product of kidney activity is urine, in which can be found more than 150 different substances.

Urine analysis allows the examination of various processes that take place inside the human organism that should show up in an external examination. As a rule, an adult's total daily amount of urine secretion is 1.2. to 1.3 liters. This depends on age, sex, nutrition, and activity. Watermelon and pumpkin consumption, for example, increases the amount of urine produced.

Urine's natural color is straw yellow. A dilute red or brown-yellow with brick-colored sediment occurs during illnesses with fever. A bloody tinge appears with increased secretion of hemoglobin: this is connected to hypertension.

Such are the external signs in a healthy person's urine and in that of someone who is sick. All the tests and indications detailed below make up a simplified system of medically monitoring trainees' health under conditions of significant physical and psychological exertions.

SOME POSSIBLE BREAKDO WNS IN HUMAN HEALTH UNDER HEAVY STRESS

One of the fundamental causes of illness is in the incorrect methods of building physical strength, leading to excessive strain and going beyond an individual's capacity for functioning. Physical training exercises have to be planned out so that the human system is not excessively fatigued by the daily regimen, and so that there are opportunities to restore the strength.

Characteristics of pathological conditions are given here.

Hypertension

This is an abrupt deterioration of health and functionality under increased physical stress. It can be acute or chronic. Acute hypertension is the result of a one-time effect of excessive stress. Chronic hypertension produces alterations that lead to disease and/or dysfunction in the various body systems.

Shock

This is a serious, common condition in those who are injured, and it is manifested in a depression of the nervous system and the functioning of all the body's systems. It is caused by extraordinary trauma, burns, poisoning, or hypertension.

First Aid

Stop the bleeding; treat the pain; apply stimulants for cardiac activity (strong tea, coffee, camphor, kordiamin). Immediately get the victim to a doctor.

Fainting

This is an instant, brief loss of consciousness. The victim goes pale, breaks into a cold sweat, and stares fixedly.

First Aid

Put the victim on his back with his feet a bit higher than his head. Then loosen his clothing, splash cold water on his face and chest, and make sure he gets fresh air.

Gravitation Shock

This is a fainting condition that can appear after physical exertion. For instance, if after a strenuous run, an individual stops for some time, he can undergo loss of consciousness. While he is running, significant dilation of the blood vessels occurs in the lower extremities along with a simultaneous constriction of the vessels in the abdominal cavity. With an abrupt halt to the run, an essential factor in the flow of blood through the veins ceases to operate: muscle contraction ("muscle pump"). This brings about a reduction in blood return, producing hypoxia in the brain. The blood's passage to the heart is constricted, blood pressure falls, and the pulse increases sharply to 190 to 200 beats per minute, bringing on loss of consciousness.

First Aid
This is the same as for fainting.

Prevention
To avoid gravitational shock, an individual should not come to a complete halt right after serious physical exertion but rather should continue moving, gradually reducing his pace and effort.

Individuals with a weak physique can experience orthostatic collapse, which is an abrupt constriction of the veins carrying blood to the heart, and it results in a standstill of the blood flow in the lower extremities and the abdominal cavity. The reason for this is weak blood vessel walls in the legs and abdominal cavity. The collapse can occur after cross-country runs, long route marches, or forced marches.

Sunstroke

This is a grave disruption of the brain's function. It is caused by prolonged activity in direct sunlight. It produces increased internal cranial pressure. It destroys body heat regulation, lowers heat exchange, and raises body temperature to 42 degrees C (sometimes 43 degrees C).

The signs are acute agitation followed by general weakness, sluggishness, nausea (sometimes with vomiting), headache, ringing ears, dizziness, nosebleeds, a drop in cardiac activity, hoarse breathing, and loss of consciousness. The face becomes covered with a fine sweat.

First Aid
Get the victim to a cool spot, arrange him in a half-lying, half-sitting position, get his clothes off, splash him with water, apply cold compresses to his head, and have him breathe spirits of ammonia.

Prevention

In hot weather, light clothing should be worn and, most important, light-complected individuals should protect their heads from direct sunlight.

Heatstroke

This is a result of disruption of body heat regulation that is caused by an intense overheating of the body, and by diminished heat dissipation. This interferes with circulation, overloads the brain's vessels with blood, and upsets heat transfer. Heatstroke can occur during strenuous physical activity in sultry, humid, high-temperature air in hot, windless, damp weather, and when wearing excessively warm clothing. The signs of and first aid for heatstroke are the same as those for sunstroke.

Drowning

This occurs when the air passages are filled with water (true drowning) and, secondly, happens during loss of consciousness in the water from hypothermia, injury, or often from panic.

In the first instance, asphyxiation (suffocation) results from an excessive buildup of carbonic acid in the blood. Respiration stops within four to five minutes, and cardiac activity within fifteen minutes. The skin takes on a bluish color. When the water is expelled from the lungs during first aid, a bloody froth often comes out of the mouth and nose.

Drowning brought on by hypothermia produces a reactive cardiac arrest caused by the cold water acting on the skin's receptor cells. The irregular, jerky movements of a drowning victim, his retention of breath (from muscle spasms of the upper respiratory tract), then shortness of breath, an increase in arterial pressure, and a higher pulse rate may give way to seizures, loss of consciousness, and subsequent cardiac arrest. In these cases the skin and mucous membranes become pale.

Causes for drowning can be found in sudden fatigue, overheating oneself in the sun, eating too much before swimming, excessive drinking, or losing consciousness from an injury upon diving into the water.

First Aid

Clear the mouth and nasal passages of any obstructing material and press the water from the lungs. Place the victim face down across your knees, and push on his back and the sides of the rib cage. Get any clothes off him and lay him on his back on a hard, level surface and perform heart massage (with the fingertips) and artificial respiration.

Heart Disease Syndrome

A syndrome is a group of symptoms/signs characterizing a particular illness.

This is characterized by pains in the heart region. This occurs when a diet is interrupted or by a prolonged route march that brings on hypertension. It is seen most often in those who have had a pathological disruption of their health, e.g., chronic tonsillitis, cholecystitis (gall bladder inflammation), a siege of angina, or grippe. Pains may appear during physical exertion, as well as during inactive periods. Medication after medical treatment and reducing stress help to relieve pain in the heart region. A systematic course of special physical exercise is also helpful in forestalling the pains.

Hypoglycemia

This is connected to an abrupt drop in blood sugar. It can occur in an individual during light physical preparation for extended exertion, during marches, tactical operations, training, forced marches, and prolonged stays in the water. A lack of carbohydrates takes a particularly heavy toll of the nervous system. The signs are sensations of extreme cold on marches, sluggishness, dizziness, agitation, a weak pulse, cold sweat, shortness of breath, paleness, disconnected speech, and notable interruption to, or full loss of, consciousness.

First Aid

Give the victim a glass of an infusion of tea or plain, strong tea along with six to eight lumps of sugar and some white bread. In cases of deep shock, get the victim to a medical establishment. For extremely long marches, halts have to be made and feeding points set up. It is recommended that two to three hours be taken for eating before extended, demanding route marches. In the interests of prevention, attention should be given to graduated physical preparation and training.

Motion Sickness

This is a pathological condition resulting from overstimulation of external physical sensors, coupled with reaction to motion by the internal organs (e.g., the stomach and the intestinal tract). In motion sickness there is an overall sensation of discomfort, manifesting itself in skin pallor, drowsiness, dizziness, nausea, increased perspiration, slowed pulse, and a fall in arterial blood pressure. The affliction occurs on rough waters (seasickness), airplane flights, in vehicle trips on dirt or paved roads with frequent curves, in ascents and descents, and especially when riding in the back of a vehicle. The combat capability and preparedness of those riding in, especially, open vehicles falls sharply. The aid to be given for this condition is in changing places in a vehicle, rest, heart stimulants (e.g., kordiamin, karvadol), and cold drinks. Prevention calls for tablets of a Dramamine-like substance.

Hypothermia

This is a general drop in body temperature with disruption of the body's functionality, up to and including a complete cessation of the latter. Hypothermia is promoted by serious exhaustion, an unwell physical condition, hunger, insufficient clothing, too much alcohol, and so on. It begins with chills, sluggishness, and tiredness. It shows itself in dozing and frequently falling asleep. Respiratory and cardiac activity is weakened, and cerebral anemia sets in.

First Aid

Take the victim to a warm place, prepare hot-water bottles, flasks of hot water, or a warm bath. Apply these and stimulate the respiratory and cardiac systems with camphor, caffeine, lobelia, hot coffee or tea, or other such substances. For loss of consciousness, use artificial respiration and heart massage.

Altitude Sickness

This results from oxygen deprivation during high-altitude activity (ascents, traverses, combat, or physical exercises). Altitude sickness' occurrence depends on a range of factors such as the peculiarity of the climate, an individual's fortitude, sex, age, physical preparedness, emotional state, speed in adjustment to altitudes, extent and level of the need for oxygen, intensity of exertion, stage of acclimatization, and so on. In general, among healthy but unprepared individuals, altitude sickness begins at heights of 2,500 to 3,000 meters and comes on a few hours after going up into the mountains.

The signs are a worsening of the general physical condition, which manifests itself in sluggishness, lack of effort, increased heartbeat, light dizziness, moderate shortness of breath during physical activity, sleepiness during the day, and insomnia at night. On the third or fourth day, the system begins to compensate and, at about 2,500 to 3,000 meters up, a feeling of euphoria appears (i.e., raised spirits, verbosity, excessive gesticulating, unreasonably good humor, and—what is dangerous—reckless behavior in the immediate surroundings). During long ascents, without acclimatization, to heights of 4,000 to 5,000 meters, the physical condition declines further and apathy sets in. Oversensitivity, impatience with others, headaches, unease, and unpleasant dreams (e.g., suffocating in one's sleep) are common. Muscular activity or movement speeds up breathing and pulse rates, producing dizziness, bodily heaviness, and head pains. Appetite falls off, nausea and vomiting may occur, and the sense of taste is distorted (e.g., cravings for acidic, sharp, and salty foods). A dryness in the throat produces a false thirst. Nosebleeds also occur.

Prevention

There has to be active acclimatization, nutritious food, psychological conditioning, and breathing oxygen with a 5-to-7 percent carbonation. An optimal and functional condition and good physical capacity are obtained with, on average, a 10-to-12-day stay at moderate altitudes. An increased capacity comes with a two-to-three-month stay at 2,000 meters.

INJURIES

Tampering with the system and rules of conducting physical training activities can bring about various injuries such as abrasions, blisters, wounds, bruises, strains and tears in soft tissue, dislocations, bone and cartilage fractures, burns, concussions, heatstroke, sunstroke, and so on.

Causes of Injuries

These stem from improper organization and methodology in carrying out activities. Related to organizational defects are the following: unreasonable schedules, too many trainees in a group (more than fifteen), and the physical unpreparedness of a group's personnel. Lack of attention to teaching principles (of continuity and gradualness), an instructor's ignorance of the body's physical capabilities in one day, the underdevelopment of trainees' individual qualities, and inferior systems of teaching safety and mutual protection are primary factors. All these can lead to training injuries (in fact, make up 40 percent of all injuries in training). As for other causes, the theory of physical education addresses violating the rules for constructing and equipping training locations (15 percent of injuries), as well as unsatisfactory instruction of training units (8 to 10 percent). Unfavorable weather conditions (e.g., cold, heat, rain, blizzards, storms) also provide causes for losses from injuries (up to 8 or 9 percent).

The uniqueness of special physical training is in its employment of methods of using strength and inflicting pain and choking in combat, dealing blows in contact karate and boxing, overcoming obstacles and positions, and jumping from airplanes and helicopters. In the organization and conduct of exercises, there has to absolutely careful thinking through of measures to prevent injuries.

In line with this, a brief description of injuries is given here.

External Damage (Cutaneous)

These are wounds, abrasions, blood blisters, callouses, burns, frostbite and chilblains, which make up nearly 35 percent of all traumas found among those engaged in physical activity.

Wounds

This is mechanically induced damage, breaking the surface of the skin or mucous membranes. Wounds may be accompanied by damage to muscles, nerves, major blood vessels, bones, and joints. Wounds can be superficial or deep—cuts, splits, tears, and so forth. The basic signs are pain, bleeding, and swelling. Wounds are dangerous because of bleeding and infection.

Bleeding

This occurs externally and internally. It can be arterial, venous, or capillary. The most dangerous is arterial: the blood spurts out in a stream, as though being thrown. In venous bleeding the blood is dark red and flows relatively slowly. Capillary bleeding has the blood coming out in small drops at the wound's surface.

First Aid: This is done by stopping the bleeding and preventing chances for infection.

There are physical, chemical, biological, and mechanical ways to stop bleeding.

• Physical means include applying cold, which contracts the vessels and lessens pain.

• Chemical means are in applying an adrenaline solution that moistens the mouth of the wound, as well as a 3-percent solution of hydrogen peroxide (which makes the blood coagulate faster).

• Biological means are only in medical facilities.

• Mechanical means are in the application of a tourniquet and raising a wounded extremity or bending at its joint to the maximum. Hands are used to press an artery against a bone and put on elastic bands or ties (as a tourniquet).

We can further examine useful mechanical means of stopping bleeding. Compresses can be used to treat capillary and moderate venous bleeding. To stop weak venous bleeding, raise the extremity. For arterial bleeding, the affected artery is pressed against the nearest bone. For heavy bleeding, a band or tie (as a tourniquet) is used. The place to apply such is on the upper third of the thigh or arm. Something soft may be placed under the tourniquet. A tourniquet should not be used for longer than two hours. In the course of one to two hours, loosen it for three to five minutes in order to allow an extremity some supply of blood and then move it higher up and retighten it. Tourniquets can be fashioned from strips of cloth, sticks, belts, scarves, or rope. If a foreign object gets into the wound, extract it with tweezers and then apply sterile gauze. Daub the edges of the wound with brilliant green. After this, apply sterile field dressings from their individual packets or sterile bandages; wrap the bandages from left to right. Bandaging is usually begun at a narrower part of the body or limb. There are many kinds of bandaging techniques: sling, figure eight, circular, radial, and so on.

Abrasions

This is a surface injury resulting from rubbing the skin against a rough surface in falls to the ground or asphalt, or down a stairway. An abrasion is accompanied by sharp pain, capillary bleeding, and secretion of lymph.

First Aid: Clean the abrasion of dirt, lay gauze pads moistened with hydrogen peroxide on it, and pat the surface dry with sterile pads. Daub the abrasion with tincture of iodine.

Blood Blisters

These are often seen among participants of long marches by foot or on skis, forced marches, and mountain training. The sign is a painful red blister on the skin. Infections can occur, bringing on fever.

First Aid: Carefully clean the skin with a piece of sterile gauze and a 3-percent solution of hydrogen peroxide, alcohol, or eau de cologne. Then it can be covered with streptocidic ointment and be bandaged.

Prevention: Properly fitted footwear, clothing, field equipment, parachute harnesses, and so on are essential. Put socks on carefully. Put adhesive tape over "future blisters."

Internal Damage (Subcutaneous)

This is tissue damage that doesn't break the skin.

Bruises

First Aid: Rest and elevate the affected part of the body. Apply cold, e.g., snow or ice (for thirteen to fifteen minutes), or cold water (up to one hour). Carefully apply chlorethyl so as not to freeze the skin. Put on a dressing. After a few days begin treatment with heat and massage.

Radiculitis

This is a malady of the radicle of the spinal cord; it is most often noted in the lower back. The causes are prolonged or excessive strain in physical activity, getting chilled, injury to the outer part of the vertebrae, an infection, or a congenital illness (which is rare). The signs are sharp pains in the lower back and limited motor activity.

Treatment: This is ameliorated by rest, dry heat to the spinal region, analgesics, and treatment by a neuropathologist.

Prevention: This is accomplished through physical conditioning, reasonable (periodical) exercising before exertions, and preventing chills.

Dislocations

These are displacements of the bones' ends from their joints or sockets, accompanied by a tearing of ligaments and sacs. Signs are acute pain, a deformation of the area of the joint because of swelling, and limited movement of the joint. In elbow dislocations, the victim is bent forward, holding the damaged arm with the other hand; in a shoulder dislocation, the victim nurses the affected arm, holding it at the elbow and keeping it away from the body.

A surgeon can repair the shoulder, using local anesthetics; but then there should be a plaster cast worn for fourteen to twenty-one days, with subsequent massage and therapy. There is related damage with ruptures and tears to the muscles, tendons, and fascia.

Treatment: The first means is to chill the affected spot and immobilize it. It is recommended that the affected area be "frozen" with chlorethyl or ice, snow, or water.

Myositis

This is an inflammation of a muscle area that results most often from hypertension or chilling. Pain may not be felt during reduction of the strain.

Treatment: This includes temporary rest, heat procedures, ointments, massage (relaxing techniques), and Dimedrol.

• • • • •

In conclusion, it should be emphasized that each exertion and each strain, especially prolonged ones, absolutely require a process of restoration of the physical system. Health has to be constantly looked after. A simple kind of monitoring can be with a daily log such as the kind kept by athletes during training. According to the system you adopt for yourself, you can make entries in the log daily, several times a week, or after extraordinary physical exertions, noting your state of health, pulse rate(s), breathing rates (before and after compulsory exercising), and measurements and data gotten from a dynamometer. In doing this, if your system is not fully recovered after significant physical and psychological exertion—say, after a two-day ski expedition with performance of different kinds of missions—look mostly for these symptoms: a quick onset of sluggishness, daytime sleepiness, light nosebleeds after blowing, and painful sensations in the joints, muscles, and tendons.

READINGS

1. Bube, H.; Feck, G.; Schtubler, H.; Trogsch, F. "Tests in Sports Practice." Moscow: *Fizkul'tura i sport*, 1968, No. 1

2. Bulochko, K.T. *Physical Training of a Reconnaissance Scout.* Moscow: Voenizdat, 1945.

3. Buchko, A.F. *Take Care of the Heart.* Stavropol'skoe knizhnoe izdanie, 1977.

4. Volkov, V.P. *A Course in Unarmed Defense—"Sambo."* Moscow: Izdanie shkol'no-kursovogo otdeleniya otdela kadrov NKVD SSSR, 1940.

5. Vaitsekhovskii, S.M. "A Trainer's Book." Moscow: *Fizkul'tura i sport*, 1971.

6. Grimak, L.P. *Psychological Training of a Parachutist.* Moscow: DOSAAF, 1966, 1971.

7. Geselevich, V.A. "Regulating Athletes' Weight." Moscow: *Fizkul'tura i sport*, 1976.

8. Geselevich, V.A. "A Medical Reference for Trainers."

Moscow: *Fizkul'tura i sport*, 1976.

9. Gulevich, D.I.; Zryagintsev, *G.I. Sambo Fighting.* Moscow: Voenizdat, 1976.

10. Gurevich, I.A. *1,500 Exercises for Modeling Training in a Circle.* Minsk: Vysshaya shkola, 1975.

11. Gurevich, I.A. *Questions in the Theory and Practice of Physical Culture and Sport.* Minsk: Vysshaya shkola, 1975.

12. Dobrovol'skii, V.K. "Prevention of Injuries, Pathological Conditions, and Falling Sick during Sports Exercises." Moscow: *Fizkul'tura i sport*, 1967.

13. "Falling Sick and Injuries during Sports Activities." Ed. Dembo, A.G. Moscow: *Meditsina*, 1970.

14. *Foreign Military Review.* Krasnaya zvezda, 1979-1985.

15. Zatsiorskii, V.M. "Physical Qualities of Athletes." Moscow: *Fizkul'tura i sport*, 1970.

16. Kabachkov, V.A. "Professional-Applied Training for Alpinists-Climbers." *Teoriya i praktika fizicheskoi kul'tury*, 1974, No. 4.

17. Koltanovskii, A.P. "General and Specialized Development Exercises." Moscow: *Fizkul'tura i sport*, 1973.

18. Letunov, S.A. Motylyanskii, R.E.; Grachevskaya, N.D. "Methods of Medical-Pedagogical Observation of Sportsmen." Moscow: *Fizkul'tura i sport*, 1965.

19. *Instruction in Physical Training in the Soviet Army and Navy.* Moscow: Voenizdat, 1979.

20. *Normative Requirements of the Soviet System of Physical Education*. Moscow: VNIIFK, 1976.

21. Ozolin, N.G. *A Contemporary System of Sports Training*. Moscow: 1971.

22. *Means of Attack and Defense*. Moscow: Voenizdat, 1959.

23. *Applied Karate*. Moscow: TsS Dinamo, 1980.

24. *Sambo Program* (40-hour and 120-hour). Moscow: USO TsS VFSO Dinamo, 1952.

25. Rodnov, V. *A Linearly Taught Program of Hand-to-Hand Combat*. Moscow: vol./No.33965, 1977.

26. Rodnov, V. *Hand-to-Hand Combat*. Moscow: 1984.

27. "Sports Massage," Ed. Makarov, V.A. Moscow: *Fizkul'tura i sport*, 1975.

28. Sukhotskii, V.I. *Physical Training of the U.S. Army*. Moscow: Voenizdat, 1970.

29. "Physiological Characteristics and Methods of Developing Endurance in Sports." Ed. Zimkin, N.V. Moscow: *Fizkul'tura i sport*, 1973.

30. "Human Physiology." Ed. Zimkin, A.V. Moscow: *Fizkul'tura i sport*, 1975.

31. Kharlampiev, A.A. "Sambo Combat." Moscow: *Fizkul'tura i sport*, 1953.

32. Chikhachev, Y.T. *Hand-to-Hand Combat*. Leningrad: Voenizdat, 1979.

33. Chumakov, E.M.; Karyakin, B.P. "Karate Yesterday and Today." Moscow: Sportirnaya bor'ba, 1972.

34. Chumakov, E.M. "Tactics of a Sambo Fighter." Moscow: *Fizkul'tura i sport*, 1976.

35. Chunin, V.V. "Structure and Content of Training Exercises Performed with a Circular Complex Configuration." *Teoriya i praktika fizicheskoi kul'tury*, 1977, No. 10.

36. Yakovlev, N.N. "A Sportsman's Education." Moscow: *Fizkul'tura i sport*, 1957.

37. Yakubovich, M.I. "Essential Defense against and Arresting a Criminal." Moscow: *Znamie*, 1976.